Developing and Designing Circular Cities:

Emerging Research and Opportunities

Elżbieta Ryńska
Warsaw University of Technology, Poland

A volume in the Practice, Progress, and Proficiency in Sustainability (PPPS) Book Series

IGI Global
DISSEMINATOR OF KNOWLEDGE

Published in the United States of America by
IGI Global
Engineering Science Reference (an imprint of IGI Global)
701 E. Chocolate Avenue
Hershey PA, USA 17033
Tel: 717-533-8845
Fax: 717-533-8661
E-mail: cust@igi-global.com
Web site: http://www.igi-global.com

Library of Congress Cataloging-in-Publication Data

Names: Rynska, Elzbieta, 1962- author.
Title: Developing and designing circular cities : emerging research and opportunities / by Elźbieta D. Ryńska.
Description: Hershey, PA : Engineering Science Reference, an imprint of IGI Global, [2020] | Includes bibliographical references and index. | Summary: "This book examines the developing and design of circular cities"-- Provided by publisher.
Identifiers: LCCN 2019035682 (print) | LCCN 2019035683 (ebook) | ISBN 9781799818861 (hardcover) | ISBN 9781799818878 (paperback) | ISBN 9781799818885 (ebook)
Subjects: LCSH: City planning. | Waste minimization. | Sustainable development. | Environmental economics.
Classification: LCC TD160 .R96 2020 (print) | LCC TD160 (ebook) | DDC 711--dc23
LC record available at https://lccn.loc.gov/2019035682
LC ebook record available at https://lccn.loc.gov/2019035683

This book is published in the IGI Global book series Practice, Progress, and Proficiency in Sustainability (PPPS) (ISSN: 2330-3271; eISSN: 2330-328X)

British Cataloguing in Publication Data
A Cataloguing in Publication record for this book is available from the British Library.

Practice, Progress, and Proficiency in Sustainability (PPPS) Book Series

Ayman Batisha
International Sustainability Institute, Egypt

ISSN:2330-3271
EISSN:2330-328X

MISSION

In a world where traditional business practices are reconsidered and economic activity is performed in a global context, new areas of economic developments are recognized as the key enablers of wealth and income production. This knowledge of information technologies provides infrastructures, systems, and services towards sustainable development.

The **Practices, Progress, and Proficiency in Sustainability (PPPS) Book Series** focuses on the local and global challenges, business opportunities, and societal needs surrounding international collaboration and sustainable development of technology. This series brings together academics, researchers, entrepreneurs, policy makers and government officers aiming to contribute to the progress and proficiency in sustainability.

COVERAGE

- Strategic Management of IT
- Innovation Networks
- Knowledge clusters
- Eco-Innovation
- Sustainable Development
- Environmental informatics
- Green Technology
- Outsourcing
- Socio-Economic
- Technological learning

IGI Global is currently accepting manuscripts for publication within this series. To submit a proposal for a volume in this series, please contact our Acquisition Editors at Acquisitions@igi-global.com or visit: http://www.igi-global.com/publish/.

The Practice, Progress, and Proficiency in Sustainability (PPPS) Book Series (ISSN 2330-3271) is published by IGI Global, 701 E. Chocolate Avenue, Hershey, PA 17033-1240, USA, www.igi-global.com. This series is composed of titles available for purchase individually; each title is edited to be contextually exclusive from any other title within the series. For pricing and ordering information please visit http://www.igi-global.com/book-series/practice-progress-proficiency-sustainability/73810. Postmaster: Send all address changes to above address.

Titles in this Series

For a list of additional titles in this series, please visit:
https://www.igi-global.com/book-series/practice-progress-proficiency-sustainability/73810

Implications of Mobility as a Service (MaaS) in Urban and Rural Environments
António Manuel Amaral (University of Minho, Portugal) Luís Barreto (ESCE-IPVC, Portugal) Sara Baltazar (ESCE-IPVC, Portugal) João Pedro Silva (IPLeiria, Portugal) and Luísa Gonçalves (IPLeiria, Portugal)
Engineering Science Reference • © 2020 • 200pp • H/C (ISBN: 9781799816140) • US $150.00

Social, Economic, and Environmental Impacts Between Sustainable Financial Systems and Financial Markets
Magdalena Ziolo (University of Szczecin, Poland)
Business Science Reference • © 2020 • 383pp • H/C (ISBN: 9781799810339) • US $245.00

Technological Innovations for Sustainability and Business Growth
Geetika Jain (UP Technical University, Lucknow, India) Harjit Singh (Amity University, Noida, India) Shahriar Akter (University of Wollongong, Australia) Alka Munjal (Amity University, Noida, India) and Harpal S. Grewal (Claflin University, USA)
Business Science Reference • © 2020 • 338pp • H/C (ISBN: 9781522599401) • US $225.00

Handbook of Research on the Conservation and Restoration of Tropical Dry Forests
Rahul Bhadouria (University of Delhi, India) Sachchidanand Tripathi (University of Delhi, India) Pratap Srivastava (University of Allahabad, India) and Pardeep Singh (University of Delhi, India)
Engineering Science Reference • © 2020 • 465pp • H/C (ISBN: 9781799800149) • US $295.00

Innovative Waste Management Technologies for Sustainable Development
Rouf Ahmad Bhat (Sri Pratap College, India) Humaira Qadri (Sri Pratap College, India & Cluster University Srinagar, India) Khursheed Ahmad Wani (Government Degree College Bijbehara, India) Gowhar Hamid Dar (Sri Pratap College, India & Cluster University Srinagar, India) and Mohammad Aneesul Mehmood (Sri Pratap College, India & Cluster University Srinagar, India)
Engineering Science Reference • © 2020 • 373pp • H/C (ISBN: 9781799800316) • US $195.00

IGI Global
DISSEMINATOR OF KNOWLEDGE

701 East Chocolate Avenue, Hershey, PA 17033, USA
Tel: 717-533-8845 x100 • Fax: 717-533-8661
E-Mail: cust@igi-global.com • www.igi-global.com

It should be noted that to most people the idea of a circular economy remains an abstract concept.

Table of Contents

Preface

The focus of this book is on the procedures which possibly should be leading to the introduction of circular economy solutions in the building construction sector perceived from the designers point of view. Implementation of circular economy solutions requires adequate education programs which, when it comes to architects and urban planners, are still too few and far apart. Most of the education is focused on traditional sustainable area studies, engineering, business and linear economy issues (Ellen Macarthur Foundation, 2017). There is more to be provided within design and innovation disciplines. It should be mentioned that Netherlands is a global leader where circular economy is concerned with TUDelft currently having the widest scope of circular teaching subjects. The Dutch focus strongly on the system thinking and design (Forlslund, Clinton, & Webster, 2018). Finland accepted a different approach with strong emphasis on a cross-sectoral collaboration. China also presents an interesting case, as already in 2006 their 11th Five Year Programme included circular economy issues embedded on all governmental levels. Circular economy is also included on a post-diploma level in the MBA Centre for Environmental Policy and the MBA in Sustainability Bard College in USA. Curricula covers the issues of the circular and sustainable development and aims to train the students in the management of chain supplies, client engagement and possible modifications according with the circular economy conditions (Goodstein, 2018). Brazilian University of Sao Paulo co-operates with the National Industrial Confederation for Circular Economy which includes stakeholders from the bio-energy, food, textile, electronic, plastics and construction sector. Together with Ellen Mc Arthur Foundation they have established four pillars of circular economy within education process. These are business models consisting of: circular design, reversible cycles, activators and systems (Ometto et al., 2018; EEA, 2017a).

Design of circular systems in manufacture processes is also visible in the Master theses presented during last few years at the Technical University of Delft – themes were created in Industrial Design Engineering Department and concerned with the circular economy and Cradle-to-cradle concept. Circular economy issues were also introduced into the curricula of the Cranfield University School of Design where teaching is based on a design thinking method and practical workshops. There are also interdisciplinary Master studies "Technology, Innovations and Management in Circular Economy" with six masterclass paths: Circularity in Practise, Biological Systems, Renewable Energy Systems, Circular Design, Circular Innovative Materials and Circular Enterprises. This University also introduced circular economy issues in other subjects. After numerous projects, it was envisaged that the subjects had to be conducted holistically and with interdisciplinary input both from the academic staff and the students (Moreno et al., 2018). There are also some emerging subjects connected with circular economy to be found in many individual schools of architecture and urban planning, and this concept is becoming of interest in more education centres and universities (Moreno et al., 2018). Education is the key issue to challenge the current linear economy perspective, which should inspire and change existing approach. There is still a problem as how to teach circular economy and which disciplines should be included (Forlslund, Clinton, & Webster, 2018). It is therefore hoped that this book will provide a support input for all those studying, or already working in the building sector. Especially those, who hope to widen the scope of their knowledge in order to provide a more sustainable future. Those who wish to find areas for possible further research can use web pages of: Ellen Mac Arthur Foundation (https://www.ellenmacarthurfoundation.org/) or Circle Cities Programme (https://www.circle-economy.com/tool/cities/) and follow research from their sites.

Circular economy is a concept widely discussed and promoted by the governments of many countries including but not limited to: EU Commission with An EU action plan for the Circular Economy (https://eur-lex.europa.eu/legal-content/EN/TXT/?uri=CELEX:52015DC0614); China with The Circular Economy in China (https://www.1421.consulting/2019/06/circular-economy-in-china/); or United States of America and Canada with such activities as The Circular Economy 100 (CE100) Network. It is not a new concept, re-use cycles have been in place since the beginnings of human civilization. Industrial revolution and the era of linear economy based on a high abundance of goods and a consumers word, to a certain level overshadowed this initial theme. Still re-use should not be treated as a choice due to poverty, but more to the awareness of creating less waste containing components which can be re-used to the benefit of the society.

Possibly one of the future directions is internationalization and globalization of the ideas, as only global approach is the correct one. This approach was initially raised in the mid half of the 20th Century, by Kenneth E. Boulding, who raised the issue of a development which relied on the unlimited input resources and output sinks. This new development was being in contrast with "closed economy" where both resources and sinks form indispensable and integral part of an economic system managed similarly to a living system. The origin of the phrase "Circular economy" may also be sought in Boulding's essay "The Economics of the Coming Spaceship Earth" (Jarrett, 1966):

The closed economy of the future might similarly be called the "spaceman" economy, in which the Earth has become a single spaceship, without unlimited reservoirs of anything, either for extraction or for pollution, and in which, therefore, man must find his place in a cyclical ecological system which is capable of continuous reproduction of material form even though it cannot escape having inputs of energy.

Already as far back as 1966, Boulding also perceived that even if our contemporary cities do not produce as much pollution as the cities of the pre-technical age, their level of influence cannot be any longer described as local. Growing urban areas have catchment area that reach to all global natural reservoirs. In 1976, EU Commission received a report "The Potential for Substituting Manpower for Energy", where Walter R. Stahel and Genevieve Reday presented an outline of the early circular economy scheme. This Report was later published as "Jobs for Tomorrow: The Potential for Substituting Manpower for Energy" (Stahel & Reday, 1981). Stahel's proposition implied that circular economy is based on natural systems, where all biological and technical nutrients can be re-introduced into a development cycle. In 1989, in article "Economics of Natural Resources and the Environment" the authors, British environmental economists Kerry Turner and David W. Pearce, pointed out that the traditional linear economy was treating environment as an unlimited waste reservoir. In turn, contemporary Circular Economy approach also includes other concepts, all sharing the idea of closed loops including, but not limited to regenerative design, biomimicry and blue economy. In 2016, Walter R. Stahel wrote in Nature magazine:

Circular economy businesses fall into two groups: those that foster reuse and extended service life through repair, remanufacture, upgrade and retrofits; and those that turn old goods into, as new resources by recycling the materials. (Rowen, 2018)

This concept is highly influenced by criteria outlined in the Hannover Principles, published in 1991, written by architect William McDonough and chemist Michael Braungart. The principles are compiled as a set of statements covering design of buildings and objects, with a forethought on their environmental impact, effect on sustainable growth and overall influence on society, and include such areas as:

- The right of humanity and nature to co-exist in a healthy, supportive, diverse and sustainable condition;
- Recognition of interdependence between natural environment and human civilisation;
- Respect towards relationships between the spirit and matter:
- Acceptance of responsibility for the consequences of design decisions by human well-being, the viability of natural systems and their right for co-existence;
- Creation of safe durable structures and objects with long-term values;
- Elimination of the concept of waste;
- Relying on the natural energy flows;
- Understanding the limitations of design;
- Seeking constant improvement by sharing knowledge.

Over the years these principles have been expanded upon and evolved, presently they present an area which is described as the Cradle to Cradle (C2C®) design protocols or standards.

As already mentioned circular economy, also called closed loop economy, is a new economic model which synthetizes such concepts as: function and service development theory (Stahel, 2006), Cradle-to-Cradle design philosophy (Brangaurt & McDonough, 2002), biomimicry (Benyus, 1998), industrial ecology (Lifset et al., 2002), natural capitalism (Lovins et al., 1999), blue economy system (Pauli, 2010). This approach uses system innovations and non-standard management procedures. It reconfigures products and services in such a way as to eliminate the issues of waste and harmful influences, uses alternative energy resources and materials, closed loops for substances and social as well as environmental capital (Ellen McArthur Foundation, 2015). Basically, it divides social and economic development from the non-renewable resources consumption level. It aspires to achieve the best-case procedures for productive tasks through efficient use of locally accessible sources, renewable and biodegradable or recycled materials. Also, maintains and enriches natural capital through control of limited supplies and sustainable use of alternative stream sources. Mentioned approach enhances efficient use of resources due to re-use of the materials within technical and biological loops and allows development of efficient systems identifying and eliminating negative external influences (reduction of waste and dangerous impacts). It has been estimated that implementation of circular economy conditions in Europe may bring gains amounting to 1.8 billion euro prior to 2030, which further means more than 0.9 milliard gains achievable under the current linear model. European GDP may grow by 11% prior 2030 and by 27% before 2050. In case of products with an average life span annual decrease in EU materials could achieve savings amounting to circa 630 milliard USD net (Ellen McArthur Foundation, 2015). Additionally, further development of the closed loop economy in Europe may reduce the CO_2 emissions from the transport, nutrition and infrastructure sectors by 48% in 2030 and by 83% before 2050 when compared to present emissions. Furthermore, it will reduce consumption of primary resources (i.e., building materials, industry, green field building sites, energy and water) by 32% in 2030 and 53% before 2050. Circular economy increases productivity and quality, also reduces transport stand still periods. It has potential to create new working places, new business services and increases innovations (Ellen Mc Arthur Foundation, 2015). Many of the circular economy conditions are already being implemented in the regional and city policies as well as the construction sector.

Therefore Chapter 1 of this book introduces the general awareness issues which emerged during the second half of 20th Century. The scope concentrates on the appearance of the circular economy approach and initial management issues pointed out by Ellen Mac Arthur Foundation. It includes general issues on Cradle-to-Cradle approach and mainly EU actions on closing the loops. With this contemporary background, the scope of this Chapter purposely skips back to the beginning of city making and initial natural approach followed by a rapid change taking place at the turn of the 20th Century. Basically, the author points out that the change in city making was mainly due to a changed management structure in construction industry. Following that, the creation of cities is shown as a bigger-economic canvas which has also underwent transformation during the last 200 years. There are two indispensable issues discussed simultaneously with circular development – these are water and air loops, both discussed in the final part of this Chapter.

Chapter 2 is dedicated to the 20th and 21st Century approaches to city making starting with the Garden City concept, through limits to growth, sustainable development and green growth. It is stressed that presently the cities are perceived as a fighting arena with duels fought mainly between economic and sustainable approach. Discussion is followed by two case studies chosen from amongst sites visited by the author and accepted as the case-best examples of buildings actually following the circular route, yet without discussing this issue.

Chapter 3 continues through to circularity in modern cities. Showing how circular approach can be applied on the level of city management and planning, as based on the Ellen Mac Arthur Foundation concept. Discussion is followed case studies for Bilbao City (Spain), Rhineland-Palatinate (Germany), and Amsterdam (Netherlands). There are other approaches as well mentioned, but for the author managerial approach is of utmost import.

Chapter 4 follows the path of circularity within Cultural Heritage stock, with emphasis on the procedure where the circular approach has to be remodeled in this particular case. As one of the possible solutions, author suggests a cultural diagnosis which should support decisions when dealing with historic substance. There are three case studies presented. Author was directly involved in the first two, and indirectly with the last one.

Chapter 5, is a research question still under discussion, concerning preparation of a circular Brief for Designers. Besides general discussions, there are two case studies formulated as Master design thesis, both being a response to this research question. Author of this book was a promoter in one case and a reviewer in the other, and found that those approaches could be later developed as an innovative scope of a Circular Brief. Chapter is finalized with closing remarks defining possible further areas of research.

LIMITATIONS AND FUTURE DIRECTIONS

Scope of this book is limited to the issue of circular loops directly associated with the idea of linear cities and buildings. The main limitation is that it does straight forward application procedures have not been formulated. Rather, the book concentrates on formulation of possible benchmarks and routes which might be taken by future research teams.

It should also be considered that within circle economy the issues of waste are one of the lowest steps which must be achieved, and much more emphasis is placed on efficient re-use of existing resources and preservation of natural environment. Therefore, waste is mentioned, as one of routes which have been followed. Since the author has already published some papers concerning this issue, those interested may refer to: Quality of Resilient Cities, the Issue of Urban Waste: Waste Management as Part of Urban Metabolism, in: Smart

Cities as a Solution for Reducing Urban Waste and Pollution, Goh Hua Bee (*eds.*), 2016, IGI Global. Above papers mainly concern construction waste and re-use of various building materials and components. Since the author is European, most of the discussion, legal issues and studies deal with EU directives and show European case studies. Nevertheless, as the author has visited many interesting sites, some of the chosen case studies also refer to other continents when the managerial route complies with circularity.

This book is also limited by the date of submittal to IGI Publishing House – this being early Fall 2019.

As for future directions – when quoting European Environmental Agency (EEA, 2019c), it may be said that "European circular economy is still in its infancy". This may be applied where urban areas and design and construction processes are concerned. Existing European initiatives, are still at a very early phase and could benefit from lighthouse investments. Change will not take place overnight, especially due to the fact that linear economy has a lot in common with consumers world and human preference lies in expansion not reduction. Whereas circularity does not mean reduction, it does mean that the approach to many areas of everyday life has to be different. One of the issues is the requirement to monitor progress on implemented circular solutions. Relevant data on production and product lifecycles are not available in information systems. Also according with EEA, circular economy policies require better integration with bio-economy and climate policies – both areas being outside the scope of this book.

REFERENCES

Benyus, J. M. (1998). *Biomimicry, Innovation inspired by Nature*. Quill Editions, HarperCollins Publishers.

Braungart, M., & McDonough, W. (2002). *Cradle to cradle: Remaking the way we make things*. Vintage.

EEA. (2017a). *Circular by design – Products in the circular economy*. EEA Report no 6/2017, EEA. Retrieved from www.eea.europa.eu.

EEA. (2019c). *Paving the way for circular economy: insights on status and potential.* EEA Report no 11/2019, EEA. Retrieved from www.eea.europa.eu

Ellen Macarthur Foundation. (2015). *Growth within a Circular Economy, Vision for a competitive Europe*. Retrieved from https://www.ellenmacarthurfoundation.org/assets/downloads/publications/EllenMacArthurFoundation_Growth-Within_July15.pdf

Ellen Macarthur Foundation. (2017). *Higher Education Programme Version 2. Circular Economy and curriculum development in higher education. Briefing notes, support & illustrative resources*. Retrieved from www.ellenmacarthurfoundation.org

Forlslund, T., Clinton, N., & Webster, K. (2018). *A global snapshot of circular economy learning offerings in higher education*. Ellen MacArthur Foundation. Retrieved from www.ellenmacarthur.com

Goodstein, E. (2018). Circular Economy in the Context of an MBA in Sustainability, Circular Economy disruptions, past, present and future. In *International Symposium Abstracts 2018*. Ellen McArthur Foundation. Retrieved from https://www.ellenmacarthurfoundation.org/assets/downloads/Circular-Economy-Symposium-Extracts-June-2018.pdf

Jarrett, H. (Ed.). (1966). *Environmental Quality in a Growing Economy*. Baltimore, MD: Resources for the Future/Johns Hopkins University Press.

Lifset, R., & Graedel, T. E. (2002). Industrial Ecology: Goals and Definitions. In *Handbook for Industrial Ecology*. Brookfield: Edward Elgar. doi:10.4337/9781843765479.00009

Moreno, M., Chamley, F., Villa, R., & Encinas, A. (2018). Implementing Circular Curriculum: Are reflection on challenges and progress in postgraduate education. In *Circular Economy disruptions, past, present and future, International Symposium Abstracts 2018*. Ellen McArthur Foundation. Retrieved from https://www.ellenmacarthurfoundation.org/assets/downloads/Circular-Economy-Symposium-Extracts-June-2018.pdf

Ometto. (2018). The collaboration of University of Sao Paulo and national Industry Confederation for Circular Economy Innovation in Brazil: from cases to systemic changes. In *Circular Economy disruptions, past, present and future, International Symposium Abstracts 2018*. Ellen McArthur Foundation. Retrieved from https://www.ellenmacarthurfoundation.org/assets/downloads/Circular-Economy-Symposium-Extracts-June-2018.pdf

Pauli, G. (2010). *Blue economy*. Taos, NM: Paradigm Publications.

Rowen, B. (2018). The circular economy concept, explained. *Government Europa Quarterly, 27*. Retrieved from http://edition.pagesuite-professional.co.uk/html5/reader/production/default.aspx?Pubname=&edid=7122fbff-5a4c-4dcb-873f-bee2e

Stahel, W. (2006). *Performance economy. Palgrave Macmillan.*

Stahel, W. R., & Reday, G. (1981). *Jobs for Tomorrow: The Potential for Substituting Manpower for Energy.* Vantage Press. Retrieved from https://www.researchgate.net/publication/40935606_Jobs_for_tomorrow_the_potential_for_substituting_manpower_for_energy

Chapter 1
Natural Approach to Circularity in Creation of Cities

ABSTRACT

This chapter shows an economic canvas introduction in part that has direct influence on the architects' and urban planning approach. It is followed by a brief explanation to the historic approach to city making, as in the past most of the urbanised areas were circular and the re-use of existing building materials was a standard issue. With the age of industrialisation and introduction of modern techniques and technologies, this attitude has changed, and the linear economic development only quickened the speed with which former solutions were forgotten. General studies showing various past approaches to urban circularity will be presented. Special attention will be paid to the sustainable city as a dynamically changing development process.

INTRODUCTION

It should be noted that the scope of knowledge of which an Architect and Urban Planner should at least be aware of, is an amalgamate of various different sciences and skills. This knowledge is born out of theory and practice and is still true since the Vitruvian times (Vitruvius, 1999). Hence, with present changes covering various development issues including building industries and changing approach as to how human surroundings should be created, designers also should insert this new approach within their professional curricula. One of the early general awareness issues has appeared during the early phase

DOI: 10.4018/978-1-7998-1886-1.ch001

of the 21st Century, and was initiated by the global economic development in many ways dependent on the growing level of energy consumption. Secondary negative effects may be seen in the Earth's atmosphere and soil pollution, extreme weather phenomenon, loss of the biodiversity, and many cases of social unrests due to reduced access to many primary goods. These changes are a challenge to the contemporary society. Unfortunately, high price of oil and gas, and unstable climatic conditions form only the visible part of a floating iceberg. Due to the rapid growth of developing countries, access to resources will be more difficult and moreover, in many cases, a lower standard of the existing natural capital (water, soil and air) will prevail. Growth of human expectations will meet a natural barrier formed by a lower accessibility to goods.

In 20th Century, early seventies introduced various technologies aiming to limit or eliminate the level of pollution emitted as a by-product during manufacturing processes. Unfortunately, such approach did not influence sourcing procedures. This phase, known also as the end-of-pipe approach, was followed by the introduction of a cleaner production system based on the introduction of technical and organization solutions allowing for a lower emissions of pollutants during manufacturing processes used together with more efficient management procedures such as: ISO 14001 or EMAS and LCA analyses. According to the main Rio Conference outcomes (1990), this particular decade was dedicated towards formulating a base for a stable and environmentally sustainable society as an alternative for a society concentrated only on growth and expansion. During the 2003 meeting in Tokyo (Kaizosha, 2003), scientists from The United Nations University described a "zero emission concept". It is assumed that this concept is the next step towards "integration of sustainable development within industrial processes" including controlling and reduction of toxic emissions and waste. These theses also have a direct connection with the sustainable environmental cycles. The main aim being total utilization of mined resources, including implementation of effective and balanced use of renewable raw materials still located in the natural environment. Waste from production processes will undergo re-use as a resource input in various new production processes. Even, if total re-use is not realistic, the phrase "zero emission level" should be discussed as a set of integrated efficient solutions used during consecutive upgrades of the cycle process. A variety of systematic approaches dealing with effective use of existing resources is already in place. These approaches include the need to preserve natural resources and mitigate the civilizations' negative influences when using the ecosystems as waste sinks. Most of them are known under a common umbrella heading of a sustainable development, later expanded into resiliency (EEA (2012c). Analytic models indicate also that with the implementation of circularity, it will be possible to double the efficient use of existing resources without lowering contemporary standard of life.

This interactive management can be followed on a diagram (Figure 1). Where each subsystem economy (i.e. building), deals with two areas concerning efficiency and eco-efficiency choices. These two areas depend on the input of fossil raw materials which after initial reuse circle back through materials and energy recycling processes to become renewable resources and be re-used within construction industry. We are used to sustainable development issues, but new changes are required to achieve this new goal. In some countries, including United States of America, the newest terminology is „industrial ecology" scoping industrial production processes created according with eco-cycles.

Research is also processed in ecologic industrial parks where effective synergies formed within individual material cycles are checked as live processes. These research sites provide data, to be later used in other sites and formulate conditions for a more efficient use of resources. Currently, quite a large number of various circular economy models dealing with

Figure 1. Circular Economy Management
Source: (EllenMcArthur, 2016)

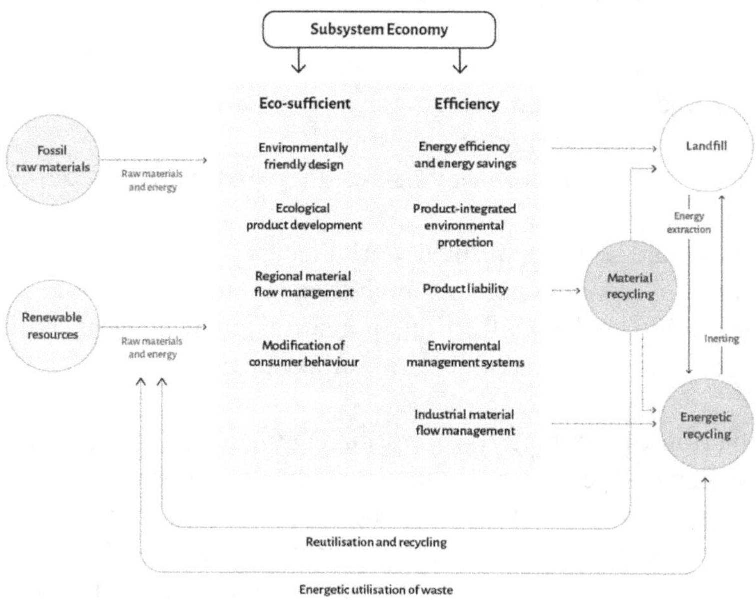

different themes is under development. Their main aim is to move away from the contemporary linear economy models. Development based on a consumption model can function properly only in the era of cheap resources and well-functioning waste sinks. Presently, only those countries which will accept circular techniques and management methods will be able to develop further. Hence, both China and Japan are implementing a 3R society (reuse, reduce and recycle) corresponding with the new economic vision. Waste reduction is the main set of themes chosen in Germany (Deilmann, 2009).

where the promoted business connections support the top-best co-operations of material loops created according to environmental solutions. Many countries and organisations are making some progress in reducing carbon emissions as part of the climate change campaign. With the growing public awareness, the inhabitants are accepting the schemes of car-sharing, lower energy use, or implementation of techniques to recycle and sort their household garbage.

In 2015, the European Commission presented a legislative circular economy package covering product's extended lifecycle from production and consumption to waste management. It was expected that these actions should benefit the environment and the economy as they are aimed at keeping both physical materials and their value as long as possible within the economic cycle, reducing waste, fostering energy savings and reducing GHG emissions. Above proposals are underpinned by 54 actions, which are currently being transformed into concrete policies (EEA, Circular Economy in Europe, 2017). The main aims of circular economy approach are described as follows:

- Preservation of natural environment through maintenance and management of sink niches.
- Lower dependence on resource suppliers.
- Cost reduction when retrieving resources ad well as required energy resources.
- Creation and keeping of local work places.
- Creation of business network including sound competition.
- Preservation and stabilization of natural environments with special emphasis on the areas of cultural importance.

Introduction of the material loop management is currently a case best solution which can be used when implementing circular economy approach. Managerial instruments include adequate sequence of tasks required during the conversion processes from the linear into circular economy. Above conditions comprise economic and ecology including integration of the social system aspects (consumption, waste management, commercial and rural areas) and industrial processes. Additional procedures concern independence between energy and material resources characteristic to each particular system indicating a need to establish network between each of participating bodies. Approach includes efficient use of resources and a higher implementation of decentralized alternative energy sourcing potential, where both private and industrial stakeholders are concerned. Management of material resources includes intelligent technologies and effective interdisciplinary planning and logistics. Proposed management system also allows activation of a major part of the existing goods potential on the micro and macroeconomic levels. In this contemporary world, where the resources are growing scarce and waste storage sinks stopped being effective, such approach also shows a potential for many new business solutions. New vast development markets dealing with effective technologies and management systems are being formed presently. Hence, contemporary requirement of water and energy sources, waste sink and energy resources call for intelligent management conditions. This approach differs from the former end-of-pipe technologies. In near future clean technologies and material flow management procedures will become the main export areas. To a certain level they are already implemented in some energy industries. Traditional linear production methods and use, are not considered sustainable anymore in comparison with other processes taking place on our Planet. It might be said that the existing model of mass production and consumption is testing the physical limits of the globe and threatening the stability of our future (Esposito et al. 2018). Even within circular economy approaches differ. Research provided within Ellen Mac Arthur Foundation focuses on design principles throughout products entire life. Other scholars focus on waste as a commodity for future use (Lacy and Rutqvist 2015). An interesting division has been formulated by Lacy and Rutqvist, where waste has been classified in four categories named as waste possibilities. These being:

- **Waste Resources**: Materials and energy that are consumed and gone when used, with fuel given as an example;
- **Products With Wasted Lifecycles**: Products that have short working lives i.e. smartphones.
- **Products With Wasted Capability**: Products used for a small part of their lives i.e. cars.
- **Wasted Embodied Values**: Components, materials, and energy not recovered from disposed products i.e. textiles

Since author of this book is concentrating more on the Ellen Mac Arthur approach, it was found of an interest that Lacy and Rutqvist concentrate more on the losses than the management process allowing for future re-use or recycling. Disregard less of the approach, should the existing major urbanization development trends, such as urban sprawl, remain unchanged; the requirement for additional resources will grow immensely. For the city areas, this means a requirement for more sites to deliver construction waste. Hence, the linear method of production based on the use of resources and removal of waste, is considered highly wasteful. In this area, transformation conditions for sustainable cities implicate a requirement for a new approach analysis of contemporary urban planning schemes in order to eliminate such practices as land infill or wasteful incineration of resources already on the pre-planning level. Furthermore, urban solutions should sustain areas and zones where adequate re-use and recycling processes might take place. Such transformation may be achieved only with the cooperation of all stakeholders – society, private business and government sectors on all administration levels. The circular economy assumptions include the requirement to establish the initial benchmark conditions to solve the issue of the growing volume of waste and hindered access to resources. The solution lies in the introduction of interdepending chains between participants in each of the production phases, allowing achievement of a maximum economic value for every component used. It is assumed that building material production cycle, use of water sources, electronic waste and also wasteful use of nutritive products may be reconfigured to achieve lower waste volumes, whereas programming of urbanized areas should be redefined to achieve a synergy effect concerning both lower waste production and economic gains. Many cities have already undertaken this experiment in some of the industrial sectors, as part of the zero waste strategic aims. The requirement for responsible consumption and production patterns is presented in SDG12 (www.cdp.net) and covers such issues as promotion of resources and energy efficiency, sustainable infrastructure and a better quality of life. It is the major point of progress towards the development of sustainable economy, balancing the requirements of the inhabitants and of our Planet. This goal dwells on the interconnectedness of both private and public sectors and encourages adoption of sustainable practices (European Commission 2018). SDG 11 concerns sustainable cities and communities and SDG 6 points to the corporate power actions when addressing shared water challenges which are also in close connection to already mentioned SDG12. Initial benchmarks for these goals, might be already implemented by 2030.

William Mc Donough, Architect, Co-author „Cradle to Cradle: Remaking the Way We Make Things (2002) as well as the author of "Something Lived, Something Dreamed (2003) and "Positive Cities" (Scientific American, July 2017) said that:

Waste does not exist in nature, because each organism contributes to the health of the whole. A fruit tree's blossoms fall to the ground and decompose into food for other living things. Bacteria and fungi feed on the organic waste of both the tree and the animals that eat its fruit, depositing nutrients in the soil that the tree can take up and convert into growth. One organism's waste becomes food for another. Nutrients flow perpetually in regenerative, cradle to cradle cycles of birth, decay and rebirth. Waste equals food.

This quotation is a good explanation of what is expected of a circular economy scheme. Industrial development initiates the level of urbanization and growth of consumption and aggregated urbanization. Industrialization and globalization effects have a deep influence on the growth of cities around the world. Mining of resources has grown twelve times between 1900 and 2015. Within last 40 years, use of resources has tripled. It is foreseen that it will double by 2050 (Circle Economy, 2018). Growing use of resources and rapid urbanization is accompanied with a rising demand for housing, food and consumption products. According to United Nations (UN, 2005), during 1900-2015, urban pollution rose from 14% to 54%, and even further growth to 66% is foreseen by 2050. This state of things creates further negative impacts on natural environment and maintaining of adequate health and well-being standards. Quantity analysis prepared by International Resource Panel (IRP) containing global requirements for resources shows that when accepting contemporary typical business conditions, consumption of resources in the cities will approach circa 90 milliard tons in 2050. This unknown in the previous history of our civilization level of consumption causes a high volume of waste which has further negative impact on natural environment. Currently we may be consuming resources at 50% faster rate than they can be replaced (Esposito et al. 2018). Large volumes of water and energy are used both during excavation and manufacturing processes. It has been estimated that more than two thirds of the total energy produced on Earth is used within the urbanized areas, which is equivalent to circa 70% of the total CO_2 emissions. The cities appear to be the major source of waste production. According to the analysis prepared by the World Bank, cities generate 1.3 milliard tons of solid waste which in 2012 was equivalent to 1.2 kg/person/

day (Circle Economy, 2018b). Traditional procedures for waste disposal mean land infill which in turn causes emissions of various pollutants. Hence, this is one of the benchmarks indicating the need to implement more economic manufacturing methods and consumption procedures in various development areas, which should at least lead to a lower volume of waste from urban areas.

Circular approach aims at a more efficient resource use through new management, creating a division between further economic development and the level of newly acquired resources. The aim is to create a foothold for a more sustainable future, allowing sustenance of some of the natural resources and screening from the potential negative effects of industrial waste. The cities should support this change as it will be of benefit when the climatic changes will become even more evident than nowadays. Some of the climatic problems and solutions are already here. In some areas like Spain, it has become too hot for the inhabitants' comfort. It is predicted that the temperature in Seville will rise to 55°C during the summer months. Therefore, in order to revive cultural street life, the Spanish government has proposed to provide external air condition system using cool underground water streams. Cartuja Qanat Project (www.cartujaqanat.com) aims to achieve local reduction of temperatures by 10°C with the use of solutions similar to Medieval Persian qanats. In a more moderate climatic zone, the city of Amsterdam is suffering from overheating and high precipitation. Here, it was decided to create a chain of green roofs. Resilio (Resilience nEtwork of Smart Innovative cLimate-adaptive rOoftops www.uia-initiative.eu) – is a network of over 10,000m^2 of intelligent roofs covered with endemic plants, with water retention process monitored so that the outcomes may be used in other locations. A different solution can be found in Copenhagen, where the city officials decided to create a Loop City (www.urbact.eu) scoping 10 small cities aiming at the 100% use of either pedestrian, mass or alternative means of transport, including autonomous vehicles. This idea was initially established as a mechanism to develop a light railway with a better transport access to the area. More and more cities promote multimodal transport, where inhabitants use various types of transport chosen as the best case sustainable solution. For example, Ghent is preparing Traffic Management as a Service platform (TMaaS – www.drive.tmass.eu) supporting mass and shared transport (bikes and cars). This development of zero-emission transport (an integral part of the circular approach scheme) is closely associated with the smog problems also presently common in many major cities. Circular approach can be also found on a more detailed level with Super Circular Estate in Kerkrade (www.uia-intiative.eu), (Oorshot and Ritzen, 2018) first European housing development 100% constructed from elements and components harvested from dismantled social buildings initially constructed in the -60-ties of the 20th Century. This solution will allow reducing the emission of CO_2 by 800 tons. This first European solution is based in a sustainable shrinkage area in Stadsregio Parkstadt Limburg in the South of Netherlands. Circularity can also be used as a guide for retrofit strategies when building upcycle processes, and as a driver towards reaching local climate goals and reduction of carbon emissions. Such approach has been chosen for the city of New York, where it is expected that 90% of contemporary buildings will still exist in thirty years' time (Linear to circular, State of Green, 2018), as well as in Copenhagen with the plan to become carbon neutral in 2025. Both cities have joined in a project exploring the use of circular design principles in architecture, engineering, and construction industry. The Circularity Lab seeks to illustrate the state-of-the-art solutions related to circular building products, materials tracking, and design for disassembly, including construction of prototype buildings in the San Francisco Bay Area. Furthermore, in 2017, the Central Denmark Region launched Circularity City Project (www.circularitycity.dk) aiming at closing the loop between supply and demand through circular solutions by engaging all involved stakeholders. In the same year, the American Institute of Architects New York, with support from the Rockefeller Foundation, published the Zero Waste Design Guidelines - 0neNYC (www.zerowastedesign.org) which address the zero waste to landfills by 2030, management of building materials and efficient water use designed into buildings and public spaces. The principles were authored by Clare Miflin, Juliette Spertus, Benjamin Miller and Christina Grace. Hence, it may be perceived that circular economy assumptions could offer the construction sector a new transformation boost when implementing modular design for disassembly solutions or prefabricated and off-site construction, designing from out-waste and recycled materials.

The idea of closing the loops shown on Figure 2, with the use of innovation and extracted resources appearing to form the input into the urban planning schemes. Proposed management issues including leading re-use, repair and remanufacture processes back to innovation arch. The city output remains as the harvesting of city resource losses (Agudelo-Vera et al 2012).

"*The transition to a circular economy is a systemic change. In addition to targeted actions affecting each phase of the value chain and key sectors, it is necessary to create the conditions under which a circular economy can flourish and resources can be mobilised.*" "Closing the loop - An EU action plan for the Circular Economy", European Commission (Closing the loop, 2015)

Figure 2. Schemes providing potential areas to close loops
Source: (Linear to circular p. 6, State of Green, 2018)

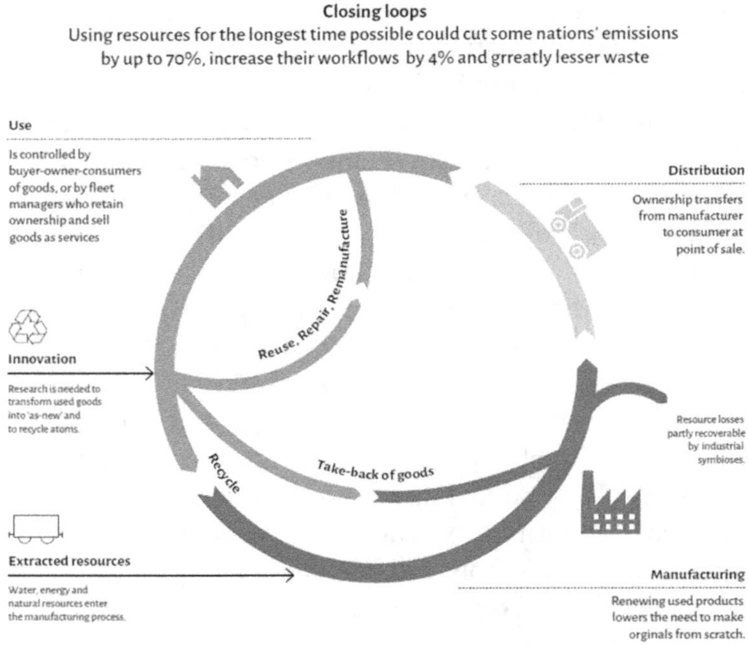

One of the loops indirectly associated with the main theme of this manuscript, and which should be mentioned, is a circular fashion economy, where presently circa half of the global clothing production is discarded within less than a year. In 2017 global fashion brands such as Adidas, Bestseller and Eileen Fisher (current signatories represent currently 7.5 percent of the global fashion market) signed a 2020 Circular Fashion System Commitment aiming also at implementation of design strategies for recyclability, and growth of the share of garments made from recycled post-consumer textile fibres (Linear to circular, State of Green, 2018). In a way, it is a comeback to the high-quality carefully selected materials allowing longer usage and several re-use cycles. This approach will reduce the fashion industry environmental footprint, since presently most clothing items end up in landfills within a year from their purchase, and only up to 12% is recycled for lower end use as e.g. insulation materials, textile boards or felts produced by some enterprises i.e. Danish Really, who in cooperation with Max Lamb presented a series of innovative benches at an International 2017 Furniture Fair (Linear to circular, State of Green, 2018). This exemplary approach should be applied where building materials and components are concerned. A new interesting approach to fashion can be found in scientific work of Seyong and Hashida (Seyong et al. 2019), where the authors analyze the relationship between the object and the user, and propose support of the clothes value by having them change users autonomously.

The Making of Cities: A Brief on Main Development Directions

Environmental, geographic and landscape conditions – in general known as location in which a city is founded, created and developed always had, and still have, great importance to those who create cities to fulfil their expectations. City evolution very much depended on the environmental, climatic and geographic parameters, perceived both as faults and assets by their inhabitants. Changing functions, various types of urban tissue, landscape and living standards were always the major issues which correspond with the relations between city inhabitants and surroundings. This approach may be described as a natural approach where urban planning and construction of particular buildings to a large extent was limited by nature and on-site existing natural capital. There is a correlation between transformation of social connections and city making procedures (Ostrowski, 1996). Therefore, it may be stated that we are presently facing a new threshold which has not been perceived as a bench mark in the years prior to industrial evolution.

When and where did the first cities appear? The question is difficult and the response varies, depending on the level of present archaeological discoveries. Initial thesis was that the first cities were founded in the fertile valleys of rivers such as Tiger, Euphrates or Nile, later the thesis pointed out Anatolia plain and Jericho area circa 7000 BC (Ostrowski 1996). Newer excavations show that this date should be moved to an earlier point in human history. Whatever the exact established date, appearance of each city was the outcome of lengthy socio-economic processes, mirrored in the human settlement's structural changes. What is never the less also true, is that each of those urbanized areas developed and were redeveloped as a layer upon layer forming a continuous cycle of material, water and environmental reuse. Hissarlik – known as the city of Troy is a very good example. In 1865, English archaeologist Frank Calvert excavated trial trenches in a field he had bought from a local farmer and in 1868, Heinrich Schliemann, a wealthy German businessman and archaeologist, also opened archaeological site in the area.

These excavations revealed several cities built in succession. Another well-known example is Knossos which has had a very long history of human habitation, beginning with the founding of the first Neolithic settlement circa 7000 BC. Archaeological excavations prove that Knossos has a thick Neolithic layer indicating that the site was a sequence of settlements before the most well-known Palace Period was initiated. And even then, Palace Period was not the last site re-use choice (Ostrowski, 1996). This proves that when the location of the cities corresponded with changing inhabitants' expectations, circular approach was common especially when the site and existing building materials were concerned – presently we would describe this management process as recycling or re-use. It was evident for our forbearers that loop economy was an issue of benefit, used also by rich societies and typical throughout the ages. I.e. Romans used building materials from Egyptian monuments whereas in Medieval ages building materials were often harvested from existing dilapidated ruined Roman buildings scavenged for materials or adapted to current needs. It can therefore be clearly perceived that re-use of materials was a normal procedure used all around our globe. Also in the later years when for example Greyfriars Friary in Leicester was suppressed during the Dissolutions of Monasteries in 1539, stone and timber was used to repair a nearby church (Waterson, 2019). This attitude changed in the 20th Century, when historic buildings started to be perceived more as obstacles than assets. Frequently, this implied destruction of buildings belonging to certain period or style. In England, Georgian buildings were demolished as they were reminiscent of the favourite style of regimes active in Germany and Italy during 1930-45. In many countries, Jewish synagogues were destroyed during the Second World War and later, when their function was often changed. The reasons for destruction were mainly ideological. In 1946, a new road system in Glasgow cut of downtown area from surrounding districts, with the centre to be demolished and rebuilt as high-rise slab structures (Waterson, 2019). It should be noted that closed loop economy was sometimes caused by a limited access to building materials, but also was due to economic understanding of less-waste creation. When this barrier was raised beginning of the 20th Century circular economy became a less obvious choice. It did not mean that circular concepts were not economically justified, but they became a secondary effect of the new attitude where long-term gains stopped being as important as short period gains. New approach was also due to a changed management structure where the building owner often stopped being the builder user. Innovative building techniques were used starting from the second half of the 19th Century shortened the time required for construction.

This in turn influenced costs. A side effect to this change is that the structures became less durable. Advanced type of steel and changes in cement additives allowed for a quicker binding of concrete, but were directly responsible for the fact that brick became less durable and could only be re-used at a much higher cost. Also the solution to hide part of the technical systems in the void between the slab and suspended ceiling or behind façade casing does not make repair, modernisation and re-use any easier. Throughout 20th Century, most common was the use of non-renewable materials i.e. the volume of cement used in China during 2011-2013 is actually equivalent with the volume of cement used in USA during 1901-2000 (Bukowski and Fabrycka, 2019). Actually re-use dwindled to a thin trickle of re-use of materials in historic buildings and only because of a shift in public attitudes which gathered its momentum in the seventies. This need for a change was also articulated by some journalists including Ian Nairn who was responsible for a criticism of the treatment given towards historic old towns. His opinion was published in 1995 Architectural Review titled Outrage. Nairn followed up this initial success with a more positive issue of 1956 called "Counter Attack", where he attempted to offer some solutions to the problems he had previously identified. The change of attitude also moved in as the Modern Movement was beginning to be discredited by many. Financial support was also created; and many heritage buildings were saved during the -70-ties and -80-ties. One of the earliest cases was a former dock area in Liverpool, where Victorian building materials and iron structural load bearing elements were re-used and adapted for a new scheme. Later on the same attitude was accepted for buildings with various functions including factory buildings. A fine example is renovation of Conway Mill Belfast (Northern Ireland, UK) – once a linen industrial area. Currently it houses a variety of functions including small business and retail units, an education center and community activities. This scheme was opened 2010 (Waterson, 2019).

Another interesting case, even though nearer present times, is Manufaktura (Figure 3.) with one of the biggest Polish commercial and recreational centres located in Lodz on the premises of the former factory of I. Poznanski. The four-year restoration of the former weaving mill, power plant, finishing facilities and the fire department building was the first example of industrial space revitalization on such a large scale in this country. This is a well-balanced combination of history and modern design. Presently, Manufaktura houses over 300 shops, restaurants, museums, discos, a bowling alley, climbing wall, cinema and a hotel. The main area is formed by a public square with fountains where many concerts and outdoors events take place. The revival

Figure 3. Views of Manufaktura complex after modernisation Lodz, Poland 2018
Source: Rynska E.

aimed at preserving the place's historical atmosphere and re-use of existing architectonic details including untreated red brick facades.

Economic Growth vs. Development of Urbanised Areas

It should be noted that the approach to urbanised areas and management of their development is very much subjected to the sequence of rapid changes and economic growth and understanding of the characteristic driving factors which form our surroundings. In the pre-Industrial Age rather small growth of population in comparison to the contemporary urbanized area was earlier described, circular fluctuation, was common. Industrial Revolution changed the general approach, and initiated the beginnings of economy closely integrated with the city development. In its initial phase – 17th -19th Century scientific interest in ecology appeared mainly in the classical works of Smith (1723-1970) and Ricardo (1772-1823), but the authors were mainly concerned with the development of early capitalistic ideas. In his book *An Inquiry into the Nature and Causes of the Wealth of* Nations (1778, last revision 1789), Smith based his analyses on the idea where the level of natural capital existing in each country corresponded with the limits of the possible economic development. Natural environment became a barrier for further growth. In case of Ricardo, in his *Principles of Political Economy and Taxation* (1817) the value of a commodity depended on the relative quantity of labour necessary for production, not on the level of compensation paid for that labour. Ricardo also believed that the process of economic development, which increased land use, principally benefited landowners. Hence, the barrier was cultivation of a lower quality land and therefore reduced standards of products. In general, 19th Century approach accepted that the relation between humans and the Nature was the control of the former over the other. This attitude was formulated following new discoveries within natural sciences and the assumption that human civilizations may very soon more efficiently control and modify the production processes. The first ecologic crisis in mid-20th Century, as well as growing awareness of the waste issues and degradation of natural environment, initiated a new approach where growing civilisation requirements stopped being perceived as external from our surroundings, but became set within the frames created by the existing ecosystems. Neoclassical economy which dominates in our present economic approach, discusses both waste growth as well as preservation of our environment and diminishing level of resources. The first area is analysed through the externality effects theories initiated by Henry Sidgwick (1838–1900) and Arthur C. Pigou (1877–1959) and the level of public goods. The negative spill-over effects are judged as the effect causing indirect costs paid by private persons, whereas positive externality is a difference between private and social benefits. This theory is criticised by the representatives of Keynesian economics mainly developed during the Great Depression (1929-1938), and initiated from the ideas presented by John Maynard Keynes in *The General Theory of Employment, Interest and Money*, published 1936. It advocates a controlled market economy, with special emphasis on the private sector and a requirement for an active role for the government interventions during recession periods. The main theme discusses the market approach and neoclassic microeconomic optimisation pointing out, that it does not provide effective mechanisms when dealing with the degradation processes and preservation of natural environment. The requirement is to maintain the sustainable environmental justice which can be sought by future generations. The differences between Neoclassical and Keynesian economics may also be found in the management processes of the non-renewable resources. Representatives of the first option assume that limited accessibility to resources initiates the substitution process determined by the existing market mechanisms. This theory is based on the Hotelling's law, concept resulting from analysis of the non-renewable resource management prepared by Harold Hotelling and published in the *Journal of Political Economy* in 1931. The volume of the non-renewable resource is constant. The present level of consumption proves that in future the consumption must be reduced. Additionally, existing resource rent is generated in the situation where the resource owner has an easy and free access to the required goods. Therefore, this rent is equal to the shadow value of the natural resource or natural capital, with the level of the rent rising as the non-renewable resource is being reduced. Described process stimulates the search for alternative resources and technologies. The representatives of the Keynesian economics also argue that the neoclassical substitutive cost-technology mechanism is inadequate, and point out that the use of resources should be government controlled through the implementation and general acceptance of the sustainable criteria. Main Keynesian points may be defined as the requirement not to imperil future generations' environmental justice. The issue is that the economic criteria applied where development is based on the non-renewable resources should not be managed and measured as the outcome of market solutions and gains. Regardless of the differences between those two approaches, at the beginning of the 70-ties of the 20th Century it was possible to distinguish appearance of new theories and methodologies such as mass-balance approach used widely in engineering and environmental analyses. In this case the assumption was, that both energy and matter may undergo transformations, but in both cases they either leave the system or accumulate according with their particular characteristics. Approach is used presently in many manufacturing procedures. I.e. BASF's uses this approach as contribution to the use of renewable raw materials in its integrated production system (BASF, 2019). Chosen management provides numerous stakeholder benefits including fossil resource saving, reduced greenhouse emissions and ready-made solutions for the clients. Method is applied to many of BASF products, such as super-absorbents, dispersions, plastics and intermediates resulting in biomass balanced products identical in terms of content and quality, but with a lower input of fossil resources. Inability to create both energy and matter, leads to the requirement that limited volume of natural resources used for further economic development must be respected as a fixed factor. In consequence there is a need to formulate rational management procedures for further exploitation, and initiate search for new technical solutions allowing to source mining sites with impeded access. Furthermore, neither matter nor energy can undergo total destruction and therefore waste must find a location in the existing space. This last concerns mainly communal and industrial waste volumes and their impact on the ecosystems and natural environment which have a limited sink capacity and therefore will not be able to limitlessly accumulate waste. Therefore forms a certain ecologic barrier appears, similar to the content of the first law of thermodynamics.

Another possible approach is the input-output analysis ("I-O"). It is a form of a macroeconomic analysis commonly used for estimating the influences of the positive or negative economic impacts and analysing the economic secondary and tertiary outcomes. This analysis specifically shows how industries are linked together through allocated inputs, and was originally developed by Wassily Leontief (1905–1999). Approach was later on adapted by other researchers (mainly Japanese) including those dealing with waste issues and emission of polluting substances (Nakamura and Kondo, 2009). Hence, initial research dealt with alternative Japanese waste management policies with regard to the regional concentration of land in-fill and the sorting of waste, including incineration processes. These models are still discussed as a simple matrix – when describing human impacts on natural capital resources (quality and quantity characteristics of direct ecological effects), and as an inverse matrix – when describing the influence of transformed natural systems on humans and standard of living.

As discussed in previous paragraphs, contemporary economy is developed on the canvas of various contradictory assumptions which in the late 20th Century initiated two possible routes. The first one is economization of natural resources – mainly concerned with the traditional Neoclassical approach to the environment. The presumption is that typical economic analysis instruments such as statistical and dynamic optimisation procedures, should be used when creating methods and instruments required to minimalize costs during ecologic resources management procedures; i.e. costs required for enhancement of the environmental standards versus costs of tasks required to achieve assumed aims. This route allows for a more efficient use of limited natural and human resources needed to achieve set forth ecologic aims. The alternative environmental preservation costs are reduced. The resources' input is reduced on the condition that such action is required for the realisation of other aims, which together support at least the existing civilisation level. Described theory conditions the need to maintain individual cooperating scientific disciplines dealing with environmental issues. The second route is ecologization of economy which enhances the multiscale and cumulative character of the phenomenon taking place in the overlapping areas of environment and development, as well as interactions taking place between various types of waste. This theory also accepts interrelations taking places between economy and natural systems. The assumption is that there is a need for a multidisciplinary participation of many different disciplines when dealing with ecological issues. Ecologic economists set forth following conditions (Lawn, 2016):

- Economic system is a subsystem of the social system, and this one in turn is a sub of the ecological system; each of the mentioned systems undergoes evolution and is conditioned by the existence of other ones.
- The real effects of the economic development are high entropy outputs; labour and capital are not an input but agents transforming low entropy resources; produced goods are only temporary effects of the process.
- Natural capital and manmade capital are not substitutes but complementary goods;
- The main aim of the economic development is achievement of a balanced strategy allowing for a maximum use of nett gains; in order to aspire to this status following issues should be resolved:
    - ecologic durability achieved through (currently not existing) global restrictions of the material and energy flows.
    - distributive justice achieved through taxes and financial initiatives.
    - efficient allocation of resources on the global market level.

Neoclassic economy is further mirrored in an anthropocentric philosophy, according to which biosphere is an instrument useful for the human race, and therefore it can be exploited with impunity. In contemporary economy and ecology, the relation between human race and biosphere may be best described with three approaches:

- The requirement to modify the anthropocentric attitude, without the need to have it totally discarded (i.e. moderate anthropocentrism) – where welfare is only one of the items shaping the standard of human life also depending on adequate environmental quality; this attitude is common to all sustainable development approaches, as well as environmentally determined development paths;
- The requirement to discard anthropocentrism and concentrate on biocentrism – where environment is the accepted a value in itself and man is only one of the living creatures;
- inability to modify contemporary societal approach (axiology - extreme anthropocentrism) – where humans are perceived as the "higher" beings; this attitude may be found in informatics revolution, genetic engineering bio and nano-technologies – which should allow humans to solve any problems concerned with the environmental degradation and adapt all new territories according with their requirements.

The classic contemporary economy, according to L. Robbins, is a science dealing with alternative use of various sources with limited access and volume. This idea was shaped based on the choice and limitation categories where environmental standards and development procedures also may be allocated. These assumptions state the existence of a limited volume of the natural and energy resources required for further development, and a limited volume of basic environmental components (air, water, space) with special emphasis on their standard. The scarcity issue is also mentioned in the non-economic context where clean environment and social requirements are mentioned.

This brief analysis discloses that rapid industrial development was the main cause for the appearance of an economy as one of the sciences addressing production means: labour, resources, finances and gains as well as particular economics i.e. building processes. Nevertheless, economic approach caused that some of the issues forming a base for the development of our civilisation including urbanisation and human health and wellbeing became marginalised. Whereas much attention was paid to the energy, resources such as: water, air and soil – also forming part of a natural capital, were for a long time treated as unlimited and with a general free access (Etkins et al. 2003).

Water and Air as an Indispensable in Circular City Development

Outside energy issues, water is an integral part of the circularity processes in our surroundings. In the Ancient times rivers such as Euphrates, Nile or Tiger initiated the location of the first civilisation areas, first within agricultural sphere and later as trading or religious centres. Water is required to quench thirst, for sanitary purposes and to sustain life of practically all creatures and plants residing on our Planet. Sustainable economy of urbanised lands depends also on the quality and quantity, and as unlimited access to water sources. Rapid industrial development characteristic for our present times is also associated with rapid urbanisation. In the beginning of 19th Century only 2% of our Planet was urbanised, at the turn of the 20th Century – circa 10%. In the first decade of the 21st Century the number of inhabitants in the urbanised areas approached 50% of the total human population. It is foreseen that this number will reach 60% before 2030 and 70% before 2050. Furthermore, it is foreseen that within next 35 years' human population residing in the urbanised areas will double. Most of the growth corresponds with the countries currently rapidly developing (UNHabitat, 2013), (CESR, 2011).

Unfortunately, in most cases this rapid growth is not based on adequate earlier management procedures concerned with basic services for the inhabitants. In many countries this is an informal sprawl process, often illegal and spontaneous. The need for potable water grows with the expansion of the city areas. Larger volumes of waste water are produced, and in many low developed countries are often removed directly to surrounding environment. It is foreseen that by 2025 annual water requirement in large agglomerations will reach circa 270 million of cubic meters per year in comparison to current 190 million cubic meters (Bergkamp et al. 2015). During the last decades this rapid development also overlaps with many globally observed climate changes which more than likely will be stronger in the coming years and will have high impact on the planning and development of the city areas. Not just the European continent faces such extreme phenomenon as: heat waves, deluge, extreme dry seasons or rapid air turbulences. They have direct influence on the management and maintenance on economic standards and investment conditions, adequate planning and urban modernisation processes. City areas are the lighthouses of economic standards and, if such major cities as London, Paris or Amsterdam will be placed under extreme climatic influences, this will have direct impact on the quality of life for all European inhabitants as well as efficient development of many regions. Climatic changes also have strong connections with social and demographic changes. Additionally, continuous urbanisation and growing competition for many sources including water might set forth rules where water will be considered as scarcity (Solarek et al. 2016). The importance of environmental capital must be included when shaping human living conditions, city landscape, social climate and human involvement in maintaining the ecosystem diversity. Such attitude is more than often forgotten in the urbanisation process. The balance of the human requirements and the status of environmental standards should be an integral element in civilisation's development. This approach scopes such issues as relations between invested and open areas (including green areas), and adequate approach to the hydrology issues when preparing city design and estimating the threshold benchmarks for the circular approach to city making.

European 2020 economic strategy for the whole continent mainly deals with the environmentally friendly approach. The key issue of such environmental economy is sustainable management of water sources, as only high standard and resilient ecosystems will create standards required to maintain healthy human society and existing standard of living. This is in contrast with the past decades mainly concerned with the implementation of rules allowing for cleaner city urban waters. Still, current level of cleanness is not adequate, even though according the Water Framework Directive plans all EU members were obliged to prepare River Basin Management Plans. Additional set of issues is concerned with the hydro-morphologic changes within water network introduced by human expansion which simultaneously cause destruction in various ecosystems, directly and indirectly connected with the water environment. Those changes are to a high level dependent on the management of agricultural areas, as well as inhabitants' expectations as to the energy sources, transport solutions and level of urbanisation. Hence, all water ecosystems should be perceived as one the natural resources – such as nutrition and energy – required for economic development of each particular area, or a whole continent. In order to define sustainable water ecosystems and sustain natural capital, we should integrate sustainable management of water sources with of land and energy capital management process. It is also indispensable to include ecologic flows, defining the volume of water required to sustain both their existence and human expectations. Sustainable water management is therefore, outside management of soil, energy and transport, one of the most important economic instruments maintaining equilibrium in other development sectors (EEA 2012b). It should include such concepts as the green infrastructure, and other actions allowing natural retention of water systems and forested areas, including reintroduction of swamps and retention areas for rainwater. Water is required for further development of economies and societies. These direct values are known as provisioning water functions. Water systems also fulfil other functions; they act as filters and dilute possible pollution substances, prevent flooding, act as fresh water sources and allow maintaining expected microclimatic balance and adequate biodiversity levels. This second set of values bears the name of a regulating and supporting ecosystem services and was systematised by Millennium Ecosystem Assessment (UN, 2005). There are definite limitations outside which human water usage may become a threat to the proper function of ecosystems, and hinder provision of adequate water standard (EEA, 2012a) for future users. Hence, it is of utmost importance that the good standard criteria will be considered together with estimation of ecologic flows (describing the quantities of water indispensable for all water systems. Many scientists point out that sustainable development defines the limits outside which water use for pure development purposes will cause destruction of ecosystems, as well as hinder procurement of future required water resources (EEA, 2012a). It is also extremely important that adequate quantity criteria will be considered simultaneously with the health standards of ecologic flows; quantity of water

Figure 4. Sustainable water allocations to ecosystems and competing users
Source: (European Waters, 2012)

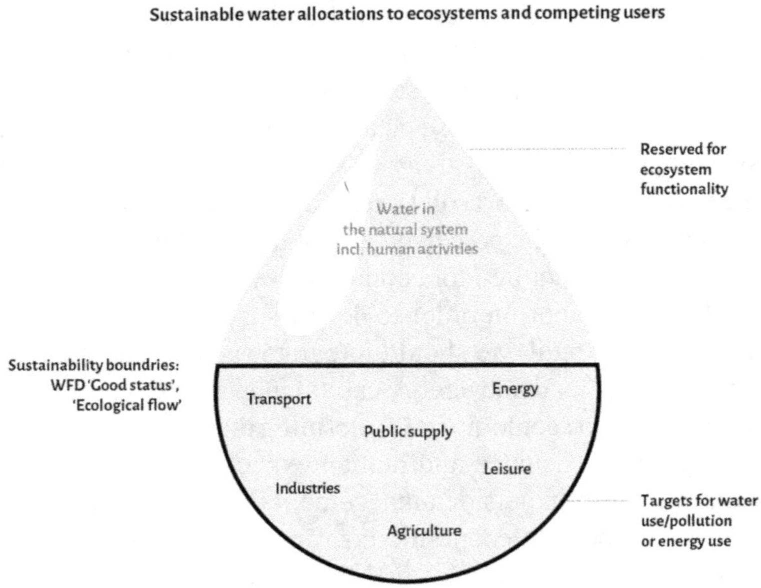

required for the water environments to function properly as well as maintain provisioning water functions for humans.

Hence, there is a requirement to define the sustainable development limits perceived as the required quantity of water sources and include water resources within circular economy loops. As many users compete for a finite volume of water resources (Figure 4), management process has to include two basic areas – civilisation needs and natural system requirements. Sustainability boundaries were discussed, but were never established as existing data is insufficient to provide a trustworthy benchmark.

Ecologic water status is a component of the biotic ("living") and abiotic ("non-living") areas of the water systems described by a set of hydro-morphologic factors, including physical shape and flow of the water masses both natural as well as created with men-made system of artificial dams, water channels or riverbed modifications. Abiotic features depend on the chemical and physical water status shaped by such factors as temperature, salinity and nutrition levels, and concentration of the pollution level of heavy metals and other chemical substances. The outcomes based on the River Basin Management Plans show that the ecologic status of European rivers is low. Provided analysis proves that:

- More than half of the open European waters (European Waters 2012) is below good standard and with low use potential; in order to achieve Water Framework Directive recommendations legal representatives, have to start on the mitigation or restauration works;
- Water courses and transboundary water have lower ecologic status and are potentially more endangered than still waters (lakes) or coastal waters;
- Most encountered water threats influencing European waters include pollution from differentiated sources causing enrichment in nutritive substances as well as hydro- morphologic threats causing changed conditions in local habitats;
- The lowest standard fresh water sources are located in central Europe, especially northern Germany, Netherlands, and in case of coastal water – Baltic Sea and North Sea areas.

Chemical status is the component of the "good water standard" which all waters should meet. In order to achieve good status, each water source must have adequate set of qualities and particular conditions defined in the Framework Water Directive (FWD). High attention should be given to medical waste having destructive influence on the human and animal hormone balance. These contaminations and unknown chemical status of circa 40% of European waters (countries outside UE) show that both monitoring conditions as well as the knowledge on possible pollution threats are inadequate. In 2011, a proposition appeared to regulate the level of contaminations defined on the priority list which may form part of the waste volume should also be included in the standards undergoing harmonization in the in the EU (EC, 2011a).

Pollution of the waters with nutritive waste such as phosphor or nitrogen sourced directly from the grey run-off waters from urbanized and agricultural areas where those chemical compounds are an integral part of industrial fertilizers. When arriving in water environment, they cause eutrophication characterized through quick development of water plants and lower level of hydrogen content. Analysis of potential threats shows that 30-35-% of the land open waters are being polluted (mainly from the water running off agricultural lands). This state concerns over 40% of the rivers and coastal waters, whereas 20-25% of their volume is also polluted by urban waste. Hence, both agricultural industry and number of inhabitants per square meter are two key issues having impact on the water pollution level. River basins surrounded by areas with more than 40% of agricultural use and inhabitant density over 100 inhabitants per square meter, are usually of low standard water quality (EEA,

2012f). Hydro-morphological pressures and natural habitats transformed through human interventions are also the main sources of negative impacts on rivers and transboundary waters (40%) as well as lakes (30%) (EEA, 2012e). Stable hydro-morphological conditions are especially important to maintain adequate ecosystem conditions such as retention and filtering, fish breeding grounds, support of bio habitats rich in bio-diversification. Changes caused by numerous human interventions, including water engineering volumes, different rainfall patterns and transformed soil covering caused by intensive urbanization, as well as construction of retention water tanks (EEA, 2012f). Therefore application of the circular economy in the areas where water is used within the manufacturing processes should benefit both urbanisation modifications and existing environment.

Outside FWD there two more equally important directives which influence biodiversity of water and marshland ecosystems. These are: Habitat Directive and Birds Directive; know under a common name of Directive Nature 2000. They scope such issues as safety, preservation and reintroduction of chosen species and ecosystems located in the areas with Nature 2000 status. Floods and droughts episodes are part of the natural hydrological and environmental cycle. During last million years species living in different European regions have become adapted to those changes forming marshlands or semidry regions with seasonal rivers. In such conditions particular ecosystems have changed adapting to the phenomenon sequence of floods and droughts. Presently, linear economic development has interrupted the existing hydrological cycle. Climatic changes in all European areas which are perceived as alternative precipitation patterns and temperature amplitudes highly influencing both the draught and flood episodes (Kossida et al. 2012) /ICM Technical Report2/2012, www.water.eionet.europa.eu_(accessed 30.10.2018). The role of climatic changes was not included in the River Basins Management Plans published in 2009; still it was one of the issues widely discussed during implementation of the strategy.

In 2012 EEA presented urban adaptation to climate change in Europe Report (EEA, 2012h) containing analysis proving that circa 20% of the European cities belonging to the more than 100,000 inhabitant's category, is located in potential flood sites. These are mainly Dutch cities, but such areas can be also found in Serbia, Slovenia, Greece and Finland. In some countries i.e. Ireland – potential threat is distributed evenly throughout the island, whereas in United Kingdom and Holland – the differences between flood endangerment vary according to chosen location. Hence, it should be noted that the cities threatened by potential floods are located in many regions.

Potential scenarios show that before 2080 circa 400,000 new inhabitants will be living within potential flood areas. Analyses also show that floods are becoming more intensive especially in northern Europe – in UK and Scandinavian Peninsula. Since 1980 at least 325 high floods have been noted and over 200 took place after year 2000. This rapid increase is caused in part through better knowledge and more efficient report making, but also it is due to changes in the use of resources. Outside river floods, such phenomenon also take place in highly urbanized zones where intensive short rainfalls may cause overflow of the sewage systems and have influence on the fresh water standards (EEA, 2012f).

Global warming analyses show intensification in the hydrological cycle and increase in the number and frequency of flood events in many European areas. More accurate estimation of those events is still not possible due to the long-term natural climatic changes and numerous artificial engineering structures located in the river systems. In the regions with lower snow precipitation and lack of snowfall during winter season, possibility of spring floods will be much lower. Insufficient information from all European countries prevents more precise scenario (EEA, 2012f). It should also be remembered that floods are also an integral part of the water and coastal ecosystems. Environmental damages caused by floods are in many ways more a consequence of human engineering than natural ecosystem changes. This further means the need to balance the social and economic damages against the values which the environmental systems might present if left in a natural state, and this includes natural floods (EEA, 2012d).

This approach also requires integration of adequate safety measures against flood impact both on the urban – city - and architectonic scale – city quarters, as well as individual buildings and industrial processes. Endangered cities may prepare various management approaches, as the strategy includes reinforcement of the building structures, decentralization of electric energy sources or underground location of all infrastructural elements. It also covers changes in the building codes, financial support of solutions imbedded in the modernization processes, implementation of various prevention solutions. Impact of floods is often international and includes areas much wider than just cities or even countries. Hence, cooperation on all administrative levels, including transboundary and international, must be included when shaping management strategies. Several adaption areas may be considered when preventing flood impact (Solarek et al. 2016):

- **"Grey" Strategies:** Concerning new buildings and infrastructure including safety measures against negative flood effects – designed with the use of innovative building solutions, economic choice of durable building materials and systematic maintenance. Higher standard of the rainwater sewage systems, provision of temporary rainwater retention areas, separation of urban sewage from the rainwater sewage lines. Innovative functional and structural solutions are often promoted – i.e. entrances to the buildings located above ground level, buildings on poles, floating buildings, temporary retention tanks, green roofs, dams and dykes; implementation of industrial processes using less water volumes or re-using water sources;
- **"Green" Strategies**: Limitations of the accepted percentage of non-impermeable ground surface areas in urban zones, maintenance of existing and introduction of new green areas, parks and gardens in the city areas. These areas also include open water areas, re-cultivation, preservation and re-introduction of marshlands, as the areas which may be used as retention tanks and water circulation areas, effective management of agricultural and forest lands, re-naturalization of river and marshland areas;
- **"Soft" Strategies**: Preparation of potential flood maps including scenarios with various climatic changes, social education, implementation of an early warning an alerting system, information containing safety measures for buildings during occurrence of high water; preparation of strategic management plans for the river basins – including building restrictions on potential flood areas, provision of retention areas for flood waters, preparation of management strategies during flood periods in connection with the rainwater systems efficiency levels, adaptation of existing technical building codes in order to include flood safety measures, introduction of financial taxes or other financial incentives i.e. definition of the lowest accepted green areas, volume of precipitation (including rainwater).

As earlier discussed both environmental and economic role of water is very complicated. Water is one of the economic resources indispensable for the civilisation and fulfils analogous role as nutrition, energy and building materials. Water is also a key component in the production and managerial system of nutrition, energy and materials. It also plays a critical role in the proper function of ecosystems, which further means that economic and environmental systems directly compete for the water sources. These issues should be evident during discussions concerning ecosystems and key resources. Simultaneously, production of energy, also from alternative sources such as bio-energies or water energy, has direct influence on the management of soils, nutrition and water resources. In order to mitigate this competition process, all man-made procedures must include water loops. These interdependencies are often described as a water-nutrition-energy loop and may allow for a synergy effect as well as provide areas of conflict between each of the components. Independencies between each management area are indicated on Figure 5.

The level of each of the stakeholders' requirement must be included when analysing possible adaption methods concerning insufficient volume of accessible water and draught periods. There are many potential conflict areas

Figure 5. The water-energy-food nexus and management influence water ecosystems and their resilience
Source: (EEA, 2012d)

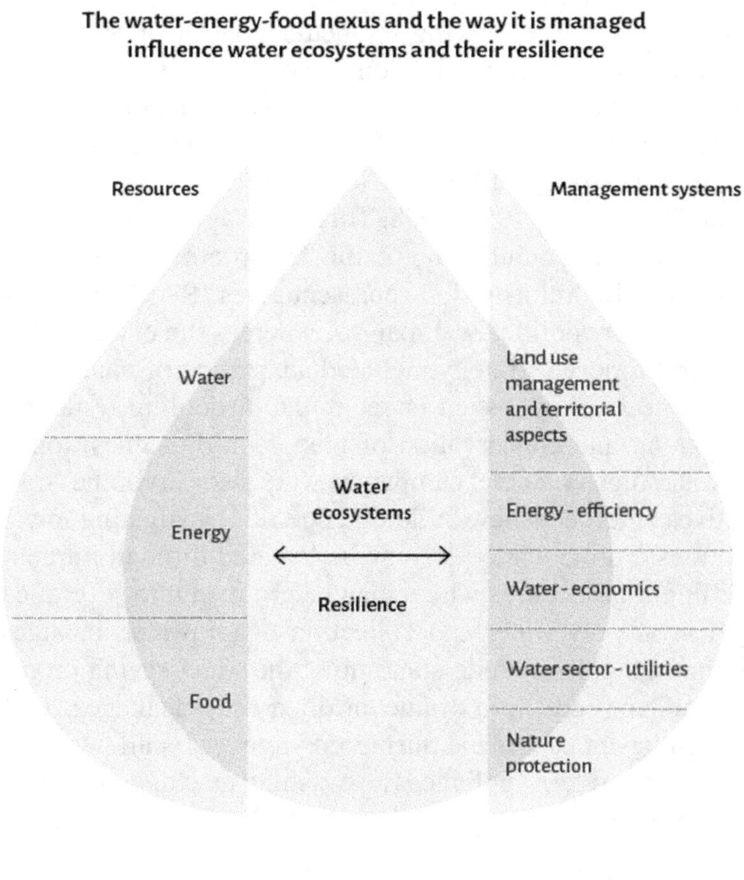

between water users, polluting waste volumes, control of legal requirements, design and introduction of responsibilities, or measurements of water usage and purchase of water licenses. These issues are still unsolved. In order to achieve the most rational use and re-use of water resources, the key issue becomes interdisciplinary co-operation between the stakeholders both on the state and business levels, and maintaining all legislative conditions. Furthermore, there is a requirement to enhance social knowledge, so that possible endangerment of water resources and undertaken solutions will be widely understood and accepted. Adaptation process to lower water usage is an universal issue, and the solutions will be possible only with the participation on all social levels.

Presently, 20% of water resources extracted in Europe is directed to the public water systems, with some minor differences between countries. Public delivery means that water is used not only in households, but also in industries, hotels, hospitals, schools and small manufacturing companies. The key component influencing the volume of used water is the number of the household inhabitants, income, and consumer behaviour as well as tourism industry. Development of various technologies including water saving appliances and solutions allowing for better control of leakage in public systems also have an important function (EEA, 2012f). One of the successful management examples is the city Saragossa located in north-western Spain. Regardless of the growing number of new inhabitants, implemented urban water management strategy allowed to lower the level of annual water use by 1600 million litres during 1995-2008. This aim was achieved through strong cooperation between Foundation for the Environment and Development (Ecodes) and local administration representatives. Reduction of water use became part of many political campaigns, whereas the quality of water also became better. Proposed strategy included adaptation of many water saving techniques and technologies such as: recirculating cooling systems and more efficient cleaning and conservation of manufacturing lines, together with individual water meters. Social campaigns were also part of the strategy. City representatives enforced a new building code rule, instigating lower level of water resources. Public water use may be lowered through implementation of many different procedures which may be grouped into several categories including: use of grey water and collection of rainwater, changes in user behaviour through general understanding of the water saving processes, use of water meters and reduction of uncontrolled water leakages. Technologic changes which have taken place during previous years allow water savings through the use of modern appliances i.e. washing machines and other kitchen and bathroom appliances. Yet, as those appliances become a standard in new buildings, there is a need to install them in existing buildings as well. Use of sanitary flush is equivalent to 25-30% of the total water use in each household. Reduction by circa 30l/day/property (Waterwise, 2010) may be achieved when using dual flush, smaller volume or a delayed filling solution. Infrared sanitary sensors and water efficient aerated shower heads and taps (Waterwise, 2010) may also be used as water saving appliances allowing for savings amounting to as much as 25 l/day/estate. Thermostatic agitators lower both the energy and volume of water required by the users. In 2011, Bio Intelligence Service (www.ec.europa.eu) also identified that water usage may be reduced in buildings by circa 10% when introducing water meters and adequate payment strategies. There is a possibility to achieve additional 5% water savings through implementation of strategic management conditions indicating use of particular products and maintaining constant monitoring of used water levels. Ready strategies may be found in certification conditions proposed by BREEAM or LEED.

Grey water section is household waste water which can be filtered, stored and later reused in toilet flushes and watering green areas (except for edible plants), reducing the volume of potable water used for other purposes. In order to lower possible microbiological contamination, either immediate use of grey water is much recommended, or additional filtration processes should be foreseen. Procedure may also be applied when collecting rainwater from roofs and other impermeable surfaces. Reduced use of potable water is important also in the countries which currently have abundance of water sources, as in such cases local sewage systems are not overburdened during heavy downpours. Tax incentives for the inhabitants who introduce proposed systems may be one of possible solutions supporting water management on a large scale. Water retaining systems may be in a different scales and complication levels, starting from the simplest home garden tank, up to a system which serves a whole city quarter. This example may be found in Berlin, where collecting rainwater from 32000m^2 of roofs is imbedded in the large scale urban solution made for the Postdamer Platz (UNEP, 2011).

When analysing efficient quantity exploitation of water resources, it should also be mentioned that the energy requirement for the manufacture and mounting processes, and maintaining of new infrastructure elements might be a source of higher GHG emissions than the traditional use of city mains. There is nevertheless a possibility for a more efficient approach when designing these solutions by lowering their carbon footprint, especially in case of water tanks and pumps (EA, 2010).

Uncontrolled leakage in the public water system is one of the major existing problems, followed by efficiency which these systems might achieve. Total elimination of leakage is practically impossible (especially due to economic and technical issues), still optimization of the management procedures leading to reduced water loss should be considered as an important task. If the level of water losses remains unchanged, it will lead to insufficient access to water sources. In case of lower pressure in the city mains, leakages might cause resource contamination through contact with surrounding environment. Quantification of the losses due to this distribution of water, including grey water, non-taxed tasks (i.e. firefighting) and illegal water use may be indirectly estimated as the difference between the volume of produced and used, measured by water meters. According to the European Benchmarking Cooperation (EBC, 2011), water system leakages create losses equal to circa $5m^3$/day/km of the water pipes. Analyses in various countries show varieties in water losses depending on the distributing firms. Presently, there are no legal restrictions where water transmission loss is concerned, except for the decisions made by some large industrial stakeholders and this in turn has direct influence on the consumer health standard and the payback period when providing maintenance of the water mains. If such calculations will neither include any external influences nor consequences of urban water mains extensions nor provide energy and material analyses, then the final outcomes will be low user quality and insufficient access to potable water sources. Obstructions in the water routes may cause infiltrations into the system and surrounding environment, causing higher local water tables and contamination of local potable water points. Water infiltration into the system causes dilution of grey waters fraction and proportional volume growth of the contamination removed into the environment. Plans to control leakages in the city mains usually cover long periods of time and they should be integrated with other tasks undertaken by the local administrative representatives i.e. transport, telecommunication, gas and heating systems. Schooling campaigns prepared for the stakeholders involved in water use during technologic process should be included as part of the management plans. Such campaigns must cover various approaches: school level education programs, information leaflets issued by local administration and water management firms, info points during public events and use of social media. Usually, when dealing with larger catchment areas, a more complicated information content should be provided.

Higher education level also include awareness on water consummation during daily tasks i.e. duration of a shower, gardening, as well as financial gains when using water efficient fittings. One of important solutions is water meter showing not only the volume of used water but also level of payments due. This last is a key issue for managers, as it allows introduction of water prices for private and business stakeholders. Water price evaluation should include: extraction, filtration and delivery costs by urban infrastructure. Energy used in this process depends on the geographical and hydrologic conditions and quality of extracted water, which will vary on location and depth of extraction. Extraction points may be located on very deep levels when the use of pumps will be required, or under riverbeds when additional filtration against pollution is needed. Energy to extract one cubic meter of water depends on location and it appears that additional data acquired in long time periods should be sourced from chosen stakeholders, as it is more reliable than comparison between stakeholders located in areas characterized by different parameters or different level of technology used. Historically, water prices in Europe only rarely mirrored total extraction costs. In effect, potable water sources were either poorly maintained or water access was restricted. Hence, present negative effects both on the environment as well as society are clearly visible. It must be noted that in fact, it is the society that pays for the cleaning process of potable water sources which have been contaminated by industry and agriculture. Changes are possible only with a wider participation of local managements and introduction of more efficient legal directives. Water Framework Directive (EU, 2000) gave an input into this better management process of water extraction costs in Europe, and article 9 contains costs reclamation procedures from polluting stakeholders. The main condition of the reclamation procedure for costs required to achieve correct potable standards is presentation of the real financial, environmental and extraction costs. This Directive points out the need for adequate contributions from different stakeholders divided at least into following areas: industry, housing and agriculture. EU members also should include social, environmental and economic estimation costs, as well as geographic and climatic parameters. The condition that purifying costs should be reclaimed from the polluting agent must include real costs based on the damages borne by the society, or based on the encroachment of accepted water standards. Motivation payment system includes implementation of a payment policy through establishment of the water effective user policies and at the same time is a contribution towards Directive's environmental goals. Already mentioned Article 9 deals only with water services, yet Article 2 defines water services as all tasks required for extraction, sequestration, preservation, purification and distribution of surface and underground water as well as all actions having high impact on the water standards. This distinction between water services and use of water sources is important when projecting the cost reclamation conditions (OECD, 2001).

Part of the research towards more effective use of water sources is Water Footprint Assessment (WFA), a quantitate approach defined as the total volume of potable water used during capital and services manufacturing processes. Initially, WFA was introduced as a simple approach indicating and calculating direct and indirect use of water resources for production and consumption (Hoekstra, 2003). Its primary aim was dissemination of knowledge to society. Later on, this approach was modified in order to estimate water quality parameters, include local issues and possible management of water extraction (mainly for agricultural purposes) (Hoekstra et al. 2011). WFA parameter distinguishes three main components: blue, green and grey. The blue component covers the volume of surface and underground waters indispensable for production means. The green one shows the scope of rainwater use retrieved before it becomes run-off water (Falkenmark, 2003). The grey component, does not concern grey waters already described in this chapter, but defines the volume of water required to naturally dilute and assimilate pollution in order to achieve required standards (Hoekstra et al. 2011). WFA coefficients are presently part of the sustainable development evaluation issues in the process of estimating environmental, social and economic factors (Hoekstra et al. 2011). Volume of water "closed" in products is also part of this analysis, WFA points out high water demand in agricultural production and large transfer of water volumes between the place of production and the place where they are used (Hoekstra and Mekonnen, 2012). In order to estimate impact level of international trade goods, all three WFA coefficients should be included (blue, green, and grey) primarily in locations where water is required and used, as well as include time and location of access to water. WFA is a good quality tool enhancing social knowledge where recycling and effective water use is concerned. In case of water extraction management criteria and estimation of real costs, provision of modelling economic software programs is also required. Such programs should help in the estimation of costs and payments, as well as estimation of quality values within the total water life cycle.

The second aspect already mentioned earlier is the air, an intricate environmental element, its standard mainly depending on the emissions from the anthropogenic changes. Important factor is the level of pollution as the major cause of premature death and disease. It is the single largest environmental health risk in major developed zones, including Europe (EEA, 2018). Heart disease and stroke are the most common reasons for premature death attributable to air pollution, followed by lung diseases and lung cancer (WHO, 2014). There is also emerging evidence that exposure to this type of pollution is associated with new-onset type 2 diabetes in adults, and it may be linked to obesity, systemic inflammation, ageing, Alzheimer's disease and dementia (RCP, 2016). Outside direct influence on living creatures, it has several important environmental impacts and may directly affect quality of water and soil and supporting ecosystem services. The volume and quality of water has a direct link with present climatic changes (EEA, 2018). Whereas, some of them contribute to the global warming, it should be added that changes in weather patterns may also may alter and magnify the transport, dispersion, deposition and formation of air pollutants in the atmosphere (EEA, 2018). Air pollution can damage building and finishing materials (i.e. corrosion of carbon steel and copper), change their properties, seriously damage cultural heritage buildings and artworks, within the social aspects lead to the loss of our history and culture. Damage includes corrosion (caused by acidifying compounds), biodegradation and soiling (caused by particles), weathering and fading of colours (caused by O_3). In fact, total effects of air pollution on health, crop and forest, ecosystems, the climate and the built environment also include the high market and non-market costs.

For past decades EU has been working on legislation which will allow to improve air quality by controlling emissions of harmful substances, improving fuel quality, and integrating environmental protection requirements into the transport, industrial and energy sectors, Circular economy idea also holds a potential for a cleaner atmosphere, as recycling and re-using processes should reduce emission of possible pollutants. Fig.6. presents a scheme of the EU clean air policy framework. It should be noted that, while most air pollutants remain stable across European Union countries, emissions of ammonia from the agricultural sector continue to rise by 0.4% from 2016 (EEA 2019b).

European Environmental Agency also pursues the requirement that all European cities need a coherent and efficient air quality measures to implement EU air quality legislation and curb air pollution. The EEA report "Europe's urban air quality – reassessing implementation challenges in cities" also identifies some of the reasons as to persistent air quality in Europe (EEA 2019 a). The Clean Air Programme for Europe (CAPE), published by the European Commission in late 2013 (European Commission, 2013), aims to ensure full compliance with existing legislation by 2020, and further improve Europe's air quality by 2030. Outside transport and industry, the main sectors contributing to emissions of air pollutants are households, energy providers and waste including landfill and waste incineration.

Figure 6. Policy framework: EU clean air policy
Source: (EEA, 2018)

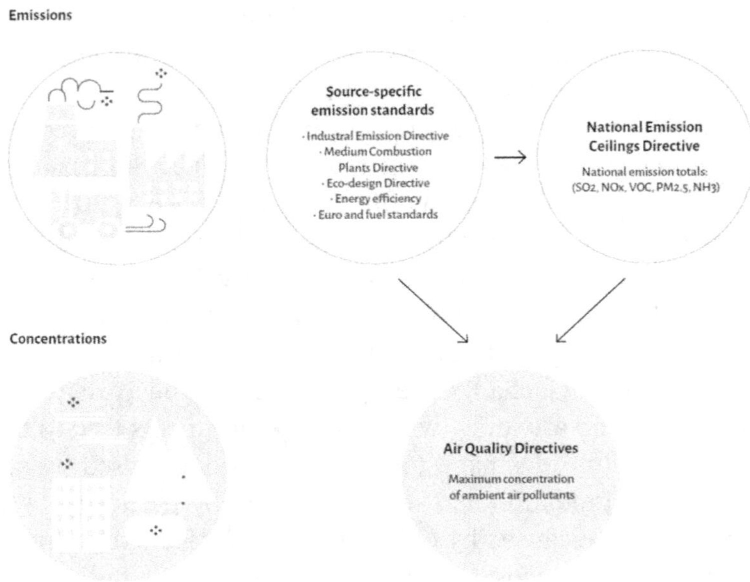

CONCLUSION

The sustainable city is a dynamically changing development process. It changes not only with the technological development but also with the growing understanding of the place which humans have within surrounding environment. One of the most important issues is presently the fact that through as our civilisation develops is strongly influences the environmental standards which are required for the human population as well as all other living creatures. Hence the air and water become the most important elements – a vital part of the natural capital - which have to be included in future development. In Europe, circular economy approach is still on the emerging level. If, and when, it is accepted and implemented it should minimise waste level, resource extraction, improve resource efficiency and contribute to biodiversity.

REFERENCES

Agudelo-Vera, C. M., Leduc, W. R., Mels, A. R., & Rijnaarts, H. M. (2012). Harvesting urban resources towards more resilient cities. *Resources, Conservation and Recycling*, *64*, 3–12. doi:10.1016/j.resconrec.2012.01.014

BASF. (n.d.). Retrieved from https://www.basf.com/global/en/who-we-are/sustainability/value-chain/renewable-raw-materials/biomass-balance.html

Bergkamp, G., Diphoorn, B., & Trommsdordd, C. (2015). *Water and development in the urban setting*. Retrieved from www.iwa_network.org/downloads/1440582594.Chapter%2010.pdf

Bukowski, H., & Fabrycka, W. (2019). *Budownictwo w obiegu zamkniętym w praktyce*. Warsaw: Polish Circular HotSpot.

CESR. (2011). *Modelling Water Scenarios and Sectoral Impacts (ClimWatAdapt)*. 2nd Stakeholder Workshop, Ministry of Rural Development, Budapest, Hungary. Retrieved from http://climwatadapt.eu/sites/default/files/2nd%20Stakeholder%20WS%20-%20Assessment%20report_Final_draft1.doc

Circle Economy. (2018). *Barriers and Recommendations to scale the circular built environment*. Retrieved from https://www.circle-economy.com/wp-content/uploads/2018/12/WBCSD_Scaling-the-circular-built-environment.pdf

Circle Economy. (2015). *Closing the Loop. EU implementation of the Circular Economy Action Plan*. Retrieved from http://ec.europa.eu/environment/circular-economy/index_en.htm

Closing the loop – An EU action plan for the Circular Economy COM/2015/0624 final. (n.d.). European Environmental Agency. Retrieved from www.eea.europa.eu

Deilmann, C. (2009). Urban Metabolism and the Surface of the City. Guiding Principles for Spatial Development in Germany. German Annual of Spatial Research and Policy, 1-16.

EA. (2010). *Energy and carbon implications of rainwater harvesting and greywater recycling*. Report SC090018, Environment Agency. Retrieved from https://assets.publishing.service.gov.uk/government/uploads/system/uploads/attachment_data/file/291745/scho0610bsmq-e-e.pdf

EBC. (2011). Retrieved from https://ec.europa.eu/info/business-economy-euro/economic-and-fiscal-policy-coordination/eu-economic-governance-monitoring-prevention-correction/european-semester/framework/europe-2020-strategy_en

EBC. (2011). Retrieved from https://eur-lex.europa.eu/legal-content/EN/TXT/?uri=CELEX%3A32000L0060

EC. (2011a). *Territorial Agenda of the European Union 2020 — Towards an Inclusive, Smart and Sustainable Europe of Diverse Regions.* Agreed at the Informal Ministerial Meeting of Ministers responsible for Spatial Planning and Territorial Development on 19th May 2011, Gödöllő, Hungary. Retrieved from http://www.eu-territorial-agenda.eu/Reference%20Documents/Final%20 TA2020.pdf

EEA. (2012a). *European Waters – current and future challenges.* Report no 9/2012. Retrieved from https://www.eea.europa.eu/publications/european-waters-synthesis-2012

EEA. (2012b). *Territorial cohesion and water management in Europe: the spatial perspective.* EEA Technical Report No 4/2012. European Environment Agency. Retrieved from http://www.eea.europa. eu/publications/territorial-cohesion-and-water-management

EEA. (2012c). *Environmental indicator report 2012 — Ecosystem resilience and resource efficiency in a green economy in Europe.* European Environment Agency. Retrieved from http://www.eea.europa.eu/publications/environmental-indicator-report-2012

EEA. (2012d). *Water resources in Europe in the context of vulnerability — EEA 2012 state of water assessment.* Report No 11/2012, European Environment Agency. Retrieved from http://www.derris.eu/wp-content/uploads/2016/03/21-water-resources-in-europe-in-the-context-of-vulnerability.pdf

EEA. (2012f). *Towards efficient use of water resources in Europe.* Report No 1/2012. Copenhagen Office for Official Publications of the European Union. Retrieved from https://www.ecologic.eu/14924

EEA. (2012h). *Urban adaptation to climate change in Europe.* Report no 2/12012. Retrieved from https://www.eea.europa.eu/publications/urban-adaptation-to-climate-change

EEA. (2017). *Circular Economy in Europe: We all have a role to play.* European Environment Agency. Retrieved from https://www.eea.europa.eu/articles/circular-economy-in-europe-we-all-have-a-roleto-play

EEA. (2018). *Air quality in Europe.* Report No 12/2018. Copenhagen Office for Official Publications of the European Union. Retrieved from https://www.eea.europa.eu/publications/air-quality-in-europe-2018

EEA. (2019a). *Challenges for achieving clean air – lessons from ten cities across Europe.* Retrieved from www.eea.europa.eu

EEA. (2019b). *Report on Ammonia Emissions from agriculture pose problems for Europe.* Retrieved from www.eea.europa.eu

EEA. (2012e). *WISE — Water Information System for Europe.* European Environment Agency. Retrieved from http://www.water.europa.eu/

Esposito, M., Tse, T., & Soufani, K. (2018). Introducing a circular economy: New Thinking with new Managerial and Policy Implications. *Berkeley Haas, California Management Review, 60*(8).

Etkins, P., Simon, S., Deutsch, I., Folke, C., & De Groot, R. (2003). A Framework for the Practical Application of the Concepts of Critical Natural Capital and Strong Sustainability. *Ecological Economics, 44*(2-3), 165-85. Retrieved from http://citeseerx.ist.psu.edu/viewdoc/download?doi=10.1.1.578.4786&rep=rep1&type=pdf

EU. (2000, December). Directive 2000/60/EC of the European Parliament and of the Council of 23 October 2000 establishing a framework for Community action in the field of water policy. *OJ L, 327*(22), 1–73.

European Commission. (2013). *The Clean Air Programme for Europe.* Retrieved from http://ec.europa.eu/environment/air/clean_air/index.htm

European Commission the High-Level Expert Group on Sustainable Finance. (2018). Financing a Sustainable European Economy. *European Commission.* Retrieved from https://ec.europa.eu/info/sites/info/files/180131-sustainable-finance-finalreport_en.pdf

European Waters – current status and future changes. (2012). Publications Office of the European Union. Retrieved from https://pdfs.semanticscholar.org/f3f3/2a4b50b0e888e3b07f3bde268d779f66d0fb.pdf

Falkenmark, M. (2003). *Water Management and Ecosystems: Living with Change*. Global Water Partnership. Retrieved from https://www.gwp.org/globalassets/global/toolbox/publications/background-papers/09-water-management-and-eco-systems.-living-with-change-2003-english.pdf

Hoekstra, A., & Mekonnen, M. (2012). *The water footprint of humanity*. Department of Water Engineering and Management, University of Twente. Retrieved from https://waterfootprint.org/media/downloads/Hoekstra-Mekonnen-2012-WaterFootprint-of-Humanity.pdf

Hoekstra, A. Y. (2003). *Virtual water trade: Proceedings of the International Expert Meeting on Virtual Water Trade, Delft, The Netherlands, 12-13 December 2002, Value of Water Research Report Series No.12, IHE, Delft, the Netherlands*. Retrieved from http://www.ayhoekstra.nl/pubs/Report12.pdf

Hoekstra, A. Y., Chapagain, A. K., Aldaya, M. M., & Mekonnen, M. M. (2011). *The Water Footprint Assessment manual: Setting the global standard*. Earthscan. Retrieved from http://www.waterfootprint.org/downloads/TheWaterFootprintAssessmentManual.pdf

Kaizosha. (2003). *Zero Emissions Manual – realizing a Zero Emissions – based on a Regional Community*. UNU.

Kossida, M., Kakava, A., Tekidou, A., Iglesias, A., & Mimikou, M. (2012). *Vulnerability to Water Scarcity and Drought in Europe. Thematic assessment for EEA Water 2012 Report*. ETC/ICM Technical Report2/2012. Retrieved from www.water.eionet.europa.eu

Lacy, P., & Rutqvist, J. (2015). *Waste to Wealth. The Circular Economy Advantage*. Palgrave Macmillan.

Lawn, P. (2016). *Resolving the Climate Change Crisis. In The Ecological Economics of Climate Change*. Springer Netherlands. doi:10.1007/978-94-017-7502-1

Linear to circular, State of Green. (2018). *Experiences from Denmark and New York on closing the loop through partnership and circular business models*. Retrieved from https://stateofgreen.com/en/uploads/2018/07/SoG_Magazine_Linear_to_circular_210x297_V05_Web-1.pdf?time=1555490701

Mekonnen, M., & Hoekstra, A. (2012). *National water footprint accounts: The green, blue and grey water footprint of production and consumption.* Institute for Water Education. Retrieved from https://research.utwente.nl/en/publications/national-water-footprint-accounts-the-green-blue-and-grey-water-f

Nakamura, S., & Kondo, Y. (2009). *Waste Input-Output Analysis. Concepts and Application to Industrial Ecology*. Springer Netherlands. doi:10.1007/978-1-4020-9902-1

OECD. (2001). *OECD Environmental Indicators. Towards Sustainable Development.* Organisation for Economic Co-operation and Development. Retrieved from https://www.oecd.org/site/worldforum/33703867.pdf

Oorshot, J., & Ritzen, M. (2018). *Super Circular Estate: urban lab for circular building practices.* Circular Economy disruptions, past, present and future, International Symposium Abstracts 2018, Ellen McArthur Foundation. Retrieved from https://www.ellenmacarthurfoundation.org/assets/downloads/Circular-Economy-Symposium-Extracts-June-2018.pdf

Ostrowski, W. (1996). *Wprowadzenie do historii budowy miast. Ludzie i Środowisko.* Warszawa: Oficyna Wydawnicza Politechniki Warszawskiej.

RCP. (2016). *Every breath we take: the lifelong impact of air pollution.* Working Party Report, Royal College of Physicians. Retrieved from https://www.rcplondon.ac.uk/projects/outputs/every-breath-we-takelifelong-impact-air-pollution

SDG12. (n.d.). Retrieved from www.cdp.net

Seyong, Y., Hasheda, T., & Ueoka, R. (2019). Rapoptosis: Renatusu via apoptosis – Prototyping using closing. In S. Yamamoto & H. Mori (Eds.), HCII 2019, LNCS 11570, (pp. 401-411). Springer Nature Switzerland AG. doi:10.1007/978-3-030-22649-7_32

Solarek, K., Rynska, E., & Mirecka, M. (2016). *Urbanistyka i architektura w zintegrowanym gospodarowaniu wodami.* Warszawa: Oficyna Wydawnicza Politechniki Warszawskiej.

UN. (2005). *The Millennium Development Goals Report.* Retrieved from https://unstats.un.org/unsd/mi/pdf/MDG%20Book.pdf

UNEP. (2011). Water: investing in natural capital. In *Towards a green economy — Pathways to sustainable development and poverty eradication.* Retrieved from http://www.unep.org/pdf/water/WAT-Water_KB_17.08_PRINT_EDITION.2011.pdf

UNHabitat. (2013). *Global Report: Planning and Design for Sustainable Urban Mobility on Human Settlements 2013.* Retrieved from https://unhabitat.org/books/planning-and-design-for-sustainable-urban-mobility-global-report-on-human-settlements-2013/

Vitruvius. (1999). Contents. In I. Rowland & T. Howe (Eds.), Vitruvius: 'Ten Books on Architecture' (pp. vii-viii). Cambridge, UK: Cambridge University Press. doi:10.1017/CBO9780511840951

Waterson, M. (2019). *Rescue & Reuse, Communities, Heritage and Architecture.* RIBA Publishing.

Waterwise. (2010). *Evidence base for large-scale water efficiency in homes — Phase II interim report.* Retrieved from http://www.waterwise.org.uk/data/resources/14/evidence-base-for-large-scale-waterefficiency-in-homes-phase-ii-interim-report.pdf

WHO. (2014). *Burden of disease from ambient air pollution for 2012 — summary of results.* World Health Organization. Retrieved from http://www.who.int/phe/health_topics/ outdoorair/databases/AAP_BoD_results_March 2014.pdf

KEY TERMS AND DEFINITIONS

Carbon Footprint: Total emissions caused by an individual, event, organization, or product, expressed as carbon dioxide equivalent – first defined in the -90-ties. Nearly 30 years later they still cannot be exactly calculated because of inadequate knowledge of and data about the complex interactions between possible contributing processes.

Entropy: A measure of the energy dispersal in the system (the higher the entropy, the higher the disorder).

GHG (Green House Emissions): A mixture of gases which absorb and emit radiant energy within the thermal infrared range. GHG cause the greenhouse effect. Carbon dioxide, methane, nitrous oxide and ozone are the main ones belonging to this category.

Great Depression: A worldwide economic depression that took place during the 1930s, starting from the United States.

Shadow Value of Natural Resource: An example of a commodity requiring shadow pricing might be the value of a view from a window indispensable to the social well-being of a community when calculating the cost of a construction project which is going to block this view. By assigning a numerical financial value to the site, economic analysts can evaluate its value to a community with regard to the costs of new construction.

Chapter 2
A Brief on the 20th and 21st Century Approaches to City Making

ABSTRACT

This chapter will delve on modern approaches to city making (eco-cities, sustainable cities, resilient cities, etc.) explaining their basics and complexity. Additionally, the demands that changing solutions place on the architects, urban planners, and other city designers will be explained. The scope should be treated as the introduction to the circular economy approach; it will also cover other development attitudes where a city was not the initial prime element even if urban planning became one of the main issues during later phases of development. Such attitudes can be traced in the mid-20th century policy making with the car transport being the leading development attitude but having a wide impact on the solutions used in most cities. It will also explain when the urbanization process became part of this economic approach. The chapter will include principles of the modern initiatives in various parts of the world and consider existing movements allowing for a more sweeping coverage.

SUSTAINABILITY AND CITY MAKING

Prior to any further discussions it should be noted that contemporary urban planning procedures concerning brown and green zones, which include sustainable development and presently circular goals, do not reduce proposed solutions to the choice of a correct location of the windmills and implementation

DOI: 10.4018/978-1-7998-1886-1.ch002

of major changes in the cultural landscape. The essential importance lies in the application of efficient and waste-less solutions giving a major positive impact on the local and global climatic changes. These actions should be followed by reduction of waste volume coming from the natural resources and lowering energy levels required for adequate function of the city environment.

The prevailing approach to a contemporary sustainable city is based on developing adaptive systems. This outlook is a combination of a triple concept: the sustainable city eco-techno-system, where the city evolves and functions similarly to a living and cyborg organisms; a balanced sustainable urban development perceived as a self-governed area managed by economic impulses derived from business priorities; an area characterized with concentration of advanced information and knowledge technologies, as well as sustainable development implementation methods (Joss, 2015).

A city perceived as a practical experiment has a much longer history since the initial ideas may be dated back as far as beginning of the 20th Century. In consequence, the naming methodology is differentiated, as it changed with the level of acquired knowledge. There is an "eco-city", a "sustainable city", a "low-energy city", a "resilient city" and a "smart city". Each name depends on the leading initiative appearing in different areas of our Globe. For example sustainable city is a seemingly simple idea which scopes both management and urban development itself. It is a trans-border and trans-cultural idea. A city which lowers the energy requirements, preserves environment, promotes urban density, reduces individual transport, lowers urban heat island effect (UHI), supports urban farming and waste recycling. Furthermore, it may be a source of potable water, extension of green areas, creation of public pedestrian friendly accessible spaces, generation of local work places, promotion of efficient use of information techniques – in short – a city ideal for inhabitants. Yet, this utopian vision is to a large extent responsible for the fact that sustainable city is still an aim to be yet reached. Hence, one of the main options is to think outside standards – in order to create, test and finally implement sustainable cities in practice. Since sustainable city is an interdisciplinary scientific area, the researchers are also trying to analyze city as an area with efficient material flows – in-takes and use of energy, water and building materials, as well as removal of waste and pollution (Puselli et al. 2004; Kennedy et al. 2007; 2011; Weisz and Steinberger, 2010; Zhang et al. 2010). Simultaneously, in case of technical disciplines, there appears a concept of a "city as an urbanised working system", where all data informing about its functions can be checked and centrally managed (Holland, 2008; Chourabi et al. 2012).

The most important issue for sustainable cities is the planning process and management. Contemporary urban planning was created and evolved as a solution to overcome the damages made by the rapid urbanisation and became an introduction of professional planning tools. At the time, this process was led under the heading of a "garden city", "new urbanism" and other similar concepts which form a foundation for the modern sustainable city making (Miller 2002, Wheeler 2000). Interdisciplinary approach also includes such disciplines as geography and landscape studies. Hence, a sustainable city exists in close relationship with a wider regional or even international context.

Transfer of knowledge between countries and continents is of utmost importance, since local implementation process will always be unique. Each city starts on a different level of development. On the most basic level, sustainable development valorises economic and social activities, which do not impact negatively on natural environment and can sustain long-term existence of future generations. This concept is in line with the knowledge that socio-economic actions which irrevocably destroy natural resources or values (water, forest, fish resources etc.) at a rate faster than they can be renewed, threaten that bio-habitats will be destroyed. This in turn will have a direct impact on the living standards of the future generations (Esposito et al. 2018).

Foundations for sustainable development may be placed in the 18th Century forest management (Grober, 2012). Yet modern definition and wide implementation in the management techniques have strong connections primarily with the Environmental and Development Commission (1987), Our Common Future (better known as the Bruntland Report, (WECD, 1987:27). Secondly, with the declaration after the Earth Summit which took place in Rio de Janeiro in 1992 (UNDESA, 1992). Standard Bruntland definition is "*development which fulfils the expectations of contemporary inhabitants of Earth without reducing development potential of the future generations*" (WECD: 1987: 27). Sustainable development evolves around three main areas of interest: environment, economy and society. There is no direct pressure on the eco-solutions, more on an economic development which includes both civilisation and preservation of natural environment. This idea has been criticised by many researchers, as there are no leading uniform formulas showing how the aims might be achieved in practice. Nevertheless, it is quite safe to state that sustainable development cannot transgress biophysical and regenerative contents of our Planet. Eco-footprint analysis, initially developed by environmental researchers such as William E. Rees and Mathis Wackernagel (Rees, 1992, Rees and Wackernagel, 1996), became one of the basic methods estimating and illustrating natural resources level of requirements set against social and economic needs. It enables to estimate quantities of resources and natural services – such as water, nutrition etc. expected through social and economic tasks; transforms the requirements into corresponding sea or landmass areas required to provide and regenerate resources, and waste sink areas. This in turn, allows calculation of an average eco-footprint of earth's inhabitant. In 2007, with the assumption of 6.7 milliard inhabitants (presently this number is established at 7.6 milliard), environmental footprint was calculated ad 1.8 global hectares/inhabitant (Ewing et al., 2010:18). Global Footprint Network also calculated that in reality eco-footprint is at least 50% larger as the present development cannot be considered sustainable. In most cases, this carbon footprint cannot be precisely assessed due to inadequate data or complex interactions, but often it is used to estimate the level of CO_2 and GHG emissions which are compared to the global carbon footprints accepted nowadays to be at a sustainable level. Hence, when considering sustainable development all three interdependent, already mentioned areas have to be included. Analysis of human civilisation and its impact on natural environment also initiates discussion on the threshold of economic development which appears to be ecologic capacity. Degradation of our surroundings also has social and cultural implications, since excessive use of natural resources by contemporary inhabitants may have negative impact on the living standards of future generations. Whereas during last 40 years, sustainable development has become a widely used and understood paradigm and a key element in many public policies, a lot of criticism is also visible. One of the negative issues is the fact that when using sustainable development in the management sector and during practical activities, the three main areas of interest are treated individually with weak interdependencies. Other critics concern actions defined as "weak" sustainable development, where natural capital i.e. forests, waters, are treated equally with the human capital (Stern, 1997; Smith et al. 2001; Etkins et al. 2003; Rydin 2010). According to this idea, environmental resources (also called eco-services), may be a part of the trading market and be exchanged for human capital substituting reduced natural capital. This idea was developed as part of the environmental ecology framework promoted by various researchers (Pearce and Turner 1990, Pearce 2002), who perceive contemporary economic models as imperfect and propose that environment also should become one of the economic factors (Pierce and Barbier, 2000). Opposite argument, is that ecological systems are unique and this aspect should be prioritised from economic and social tasks. Dying of species or excessive use of natural resources for pure economic gains may lead to environmental degradation beyond repair, including many unknown by-effects which may appear at some point in future. This argument is quoted as the requirement for a more effective sustainable development promoting importance of ecological systems. There is a need for alternative approach differing from the contemporary economic, business and social procedures.

- **The Brundtland Report**: Accepted as the key document, does not support any of the contemporary existing economic models, hence sustainable development is dominated by mediocre attitudes without any criticism towards present global economic trends. Nevertheless, cooperation between different disciplines appears to be one of the main paths. For example sustainable transport system can be achieved only through cooperation and coordination techniques between various transport systems – as this is the only way to achieve a more integrated, environmentally friendly and accessible solutions. It must be noted, that currently, cities are a place of profound struggle between economy and sustainable development.

Economic Approach to Sustainable Cities

According to 21st Century assumptions – which by many is called "urban age" (Joss, 2012) – city areas become a site for innovations for a variety of changes initiated by rapid urbanization processes. Management of sustainable areas becomes therefore not only a key issue for all city zones, but for all 21st Century (UN-Habitat, 2003). The cities should play a very special role when the climatic changes are considered. Hence, appropriate management of urban processes including environmental protection becomes a key issue where sustainable development is considered. City agglomerations are main business and industrial centres with a strong impact on the development of countries and regions. Research shows for example 64 British cities cover circa 9% of the country area, but generate circa 60% GVA, whereas annual production in metropolitan areas of New York, Chicago or Los Angeles is higher than in Sweden, Norway or Poland (Centre for Cities, 2013:2). This change in economic and political approach to cities has appeared only within last years and some researches are still unsure whether approach is correct. Even though within global economy, the cities are assumed to be individual players, it does not mean that their governments wish to act in this way (Peck and Tickell, 1994). The assumption that individual cities would gain economic strength is counterbalanced by the fact that many of the city governments chose a similar development path which include reduced access to public services and wider cooperation with corporate enterprises (Hackworth, 2007). Moreover they have moved away from management tasks to business areas (Oatley 1998: 4). Regardless of differences, it is clear that the cities with strongest economies, also have more advanced connections with the globalization processes (Brenner, 2004).

The fact that the cities as social and economic action centers relay heavily on resources, both when it comes to energy, as well as potable water, building materials and soil and therefore are also the largest pollution emitters, is the second reason why they are perceived as possible environmental players. It has been calculated that city areas emit 60-70% of total GHG (calculated based on emissions generated both during production, consumption and service processes). Furthermore, especially in Africa and Latin America, they are also a place of social inequality (UN-Habitat, 2010:10). The secondary effect is quick development of slums in countries undergoing urbanization, inhabited by circa 32% of city inhabitants (UN-Habitat, 2010: 7).

When considering prior arguments, sustainable development is possible only, if environmental, economic and social issues are solved simultaneously. In this case a city appears to have an adequate scale for interventions including transformation of exiting governmental structures. There is one more additional evidence proving close association between urbanized areas and this concerns existing rate of development. Year 2008 was crucial for human civilization, as at that date half of the human population was living either in the cities or urbanized areas (UN-Habitat, 2009b; UN-Habitat, 2011), i.e. it was estimated that in 2000-20 in China, due to intensive migration and urbanization of rural areas, circa 300 million inhabitants will become city dwellers. Hence, city management policies are not limited only to existing cities, but also concern development and preparation process required for the cities and urban areas which are will appear in near future. There are many alternative propositions concerning sustainable development. Some should be considered as utopia, others have distinctive features which might prove to be precious when considering further development of cities.

According to Joss (Joss, 2015), the key to sustainability should be sought in the concept of green cities which together with the core recommendations of the United Nations Environment Program published by the agency in *21 Issues for the 21st Century*. This report describes sustainable cities as areas with superior environmental quality and health and wellbeing conditions;

which state can be achieved through implementation of various solutions including compact and mixed use development, efficient transport, use of alternative energy sources and reduced ecological footprint. Both UN-Habitat and United Nations assume that the cities are both source and solutions of many environmental challenges (UN-Habitat, 2011), (UN, 2013).

ENVIRONMENTAL STRATEGIES

Figure 1 presents a scheme of key environmental policy narratives dating back to the beginning of 20th Century confirming that the idea as such can be sourced to the early 19th Century sanitary reforms and appearance of the Ebezener Howard's Garden City concept which seemed to be the answer to detrimental conditions characteristic to the British Victorian city centers (Wheeler and Beatley, 2009; Kargon and Molella, 2008).

Garden Cities were to provide both employment and improved living standards and cultural support, but still have a direct contact with rural countryside, as this was the understanding of natural environment of those times. Crucial to this assumption was the decentralization. Garden cities were built in many countries, but none of them fully followed Howard's concept including Japan, where it was adopted as an attempt to mitigate industrialization effects. A concept of Techno-cities followed soon after, later

Figure 1. Overview of key city policies
Source: (Diagram prepared based on Joss, 2015 pp.77)

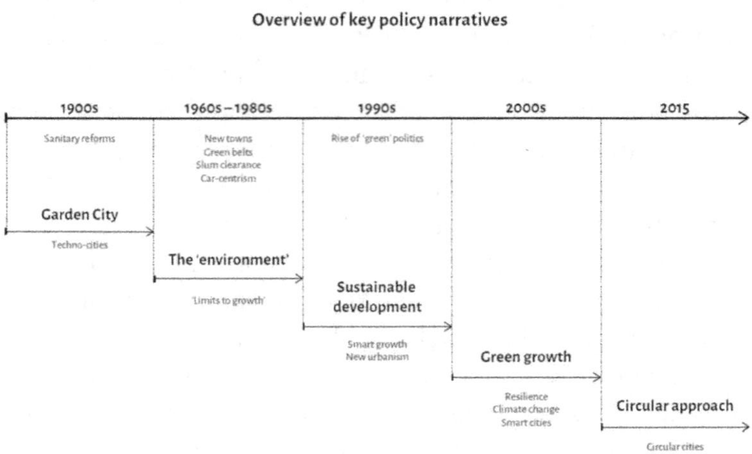

on to be supported with the first Athenian Chart (1931) and the introduction of green belts (1950-60), clearance of slums, but also introduction of a car-conditioned city divided onto mono-functional areas.

The Minamata incidents of 1956 and 1965 were the turning points for many urban management choices. The incidents appeared in a form of a neurological syndrome which was due to mercury poisoning from industrial wastewater discarded into Minamata Bay, and became an instrument to create a broad public awareness of the detrimental industrial production outcomes both for the surrounding environment and human health. Since then, industrial ecology and industrial symbiosis which deal mainly with waste reduction and recycling and relocation of businesses in order to share resources, are the main characteristic factors in Japanese eco-cities. This case was echoed throughout many countries with the most known publication *Silent Spring* written by environmental scientist Rachel Carson in 1962. That was the turning point for changing attitudes and in 1973 Denmark became the first country with official environmental protection laws. Soon it was followed by the redefinition of the city in wider environmental terms with the Urban Ecology Movement (1970) and research and innovations led by Richard Register. In his Eco-city, Berkley (Register, 1987) describes a vision complemented with a set of practical propositions concerning rebuilding of the California Berkeley (USA) as an ecologically healthy city. Idea very soon became a catalyst for a growing international movement dedicated towards sustainable urban development conditioned through research, co-operation between representatives of local administrations and implementation of practical prototypes. Movement's mission was published in 1996, where two basic conditions were described: efficient use of existing soils and development of a compact city and establishment of transit nodes. One of the main assumptions was accessibility and vicinity of public transport. These solutions touched on a much wider problem, as most of the large agglomeration face intensive often irrational housing construction sites located on the suburban outskirts forming city sprawl effect. They are often associated with transport to the city centers – so characteristic nowadays not only to American cities. Other characteristic issues such as: redevelopment of degraded areas, limitation of local pollution, efficient energy use, lower levels of CO_2 emissions – were not included as yet at this stage. One of the interesting cities which was negatively impacted by car transport is Utrecht (Netherlands), where 60 years back, due to changing economic and the requirement for a better city access the main canal area was changed to a six lane - Catharijnebaan. Fifty years later train tracks and the throughway - were a barrier that increased the perceived distance to the city Centre considerably. The restructuring of the area in order to connect the two halves of the city, increase the facilities available to the growing population of the city, to increase the allure and the safety of the area, and to create a more pleasing transition took over a planning process which lasted over 20 years, as was finalized with return of the canal water areas, less traffic in the downtown and more emphasis on mass transport (Buijze, 2013). Previously, indication that natural environment perspective is important was located more in the aesthetic area, than in preservation of natural resources was common.

Urban ecology movement also has strong social features, with special emphasis on the social inequality, support for women and minor ethnic groups or those with limited movement abilities. Once more, this movement mirrors the main features of that epoch. First assumptions for the Eco-city promoted by Jeffrey Kenworthy, are analogously concentrated on the connections between urban form and transport. The most cited paper "The Eco-city: Ten Key Transport and Planning Dimensions for Sustainable City development (Kenworthy 2006), justifies transport solutions as the key issues with direct influence on the North America and Australian development, together with high use of energy resources, large quantities of materials and waste. The ten key guidelines are the direct outcome of the transport systems and a challenge to limit the "car dependence" through adequate structure of the urban tissue, and most of all of promotion areas with alternative mixed functions. For Kenworthy, urbanized space is a key to sustainable development perceived as an area required in the city to sustain adequate living area and all economic activities. Area is applied as a critical parameter when defining the level of sustainable development especially when transport conditions are considered. In comparison with Urban Ecology, Eco-City (Zhou et al.2013) places higher emphasis on environmental friendly technologies to source energy and water, waste management and form closed system cycles to achieve better efficient performance. The aim of these technologies is to achieve a possibility that cities will be using natural regional capital and start constructing closed loop infrastructure where recycling and re-use is part of the system. This selection of tasks is sustained with a need to create a place which will simultaneously fulfill inhabitants' expectations and will be used as an interactive social space. This approach includes social debates and collective decision making.

Green Urbanism Assumption (Lehman 2010): Transformation of the city area into sustainable development zone is at the heart of the 15 conditions defining green urbanism, which have been composed around "triple zero" idea: zero use of energy resources, zero waste level and zero emissions - basically of low or of nearly zero type (Lehman, 2010:230). Conditions confirm that this approach forms a threshold between a sustainable city and a city characterised by low carbon emissions, efficient renewable energy use as well as any other resources. Hence, first three conditions mainly concern climate, renewable energy resources in order to reduce CO_2 emissions and zero waste. Main approach is that a city is treated as an energy independent zone and based on a decentralised energy system and zero-waste urban, area forming a closed circular management system. Additionally Lehman indicates that local climatic conditions (landscape and orientation, sunlight level, precipitation, humidity, prevailing wind directions etc.) characteristic to particular site, should form the core for defining and creating urban sustainable conditions. Remaining assumptions are analogous to those which can be found in other theories and concern preservation of potable water resources, sustaining bio-diversification, creation of local production and food transport chains through intensification of building density along transport focus joints. Analogously as in other cases there is an option to formulate balance (even though there is insufficient information as to how to achieve this state) through "local identification" and "a feeling of belonging to a place".

Since the beginning of the 21st Century many sustainable initiatives have already taken place or are currently being developed. Many of them concentrate on the climatic changes and process of lowering GHG emissions during construction and use of buildings. Presently a lot is also discussed on the overall reduction of resources. This approach is possibly due to the fact that the issue of climatic changes is nowadays considered as one of the gravest global problems. Reduction of carbon emissions has direct impact on the standard of air in many cities. Another vital issue is limiting dependence on individual energy suppliers, both when international and national suppliers are considered. Carbon emissions are a concrete, practical and measurable interpretation of the undefined sustainable development concept.

The city of Copenhagen is a particular case of a low carbon policy and in 2014 was awarded the European Green Capital Award[1].

In 2018, this prize was awarded to one of the oldest Dutch cities – Nijmegen, located on the River Wall just beyond the German border. After Second World War it played the role of one the largest inland ports. It also was a quickly developing industrial centre based on the incomings from coal industries which are currently being taken over by other industries. This transformation concerns not only energy sources (wind energy investments supported by (WindPower Nijmegen), but also relocation of work places into environmental technologies which are becoming the leading city working sector. Cooperation taking place between the city government, inhabitants and various business stakeholder is very impressive. It includes numerous initiatives where Nijmegen is taking the position of an eco-ambassador. Another leading set of investments is the transport sector, with numerous investments in bicycle paths and waste management. (Nijmegen. European Green Capital, 2018).

One of the newest approaches is development of resilient cities (Newman et al. 2009), a difficult term derived from medicine, often without adequate translation into other languages. Resiliency issues mainly are considered as an approach to create cities which are resilient towards negative climatic impacts. As has been mentioned earlier, cities are becoming to be treated not only as areas characterized with a high emission level of pollution, but also zones which are very sensitive where climatic changes are concerned – including global higher level of seas, floods or limited access to potable water sources. In effect, negative impact does not apply only to the building physical structure, transport or energy systems, but also human health and various support services. The most impacted areas concentrate in the coastal zone, and numerous cities located on every continent are located in such areas. Specialists employed by UN-HABITAT have estimated that 13% of the city inhabitants dwell in such fragile areas (UN-Habitat, 2011). Hence, urban planning process should also include the issue of resiliency through implementation of adequate management mechanisms concerning not only climatic safety precautions, but also social-economic issues, nutrition supplies and health requirements. Regardless of the variety of issues it is possible to discern characteristic areas such as adaptation and flexibility of accepted solutions with special attention paid to management procedures. Development of short local supply chains – being one of the safety measures from the environmental and climatic negative impacts which may take place in surrounding districts. These conditions have been illustrated as five major resiliency issues formulated by Rockefeller Foundation (Rockefeller Foundation, 2013). Described assumption touches on the natural and human caused catastrophes and includes description explaining how the city managers and inhabitants can prepare to face possible crises. Specific areas of interest concern food supplies, housing, health issues and production industries. Hence the five characteristic environmental elements are: 1. additional capacity for food storage and alternative procedures when major systems are destroyed 2. Flexibility, ability for transformation, change or adaptation in case of disasters 3. Limitation or "safe unsuccessful decisions" which will not impact other parts of the system 4. A rapid return to the state of full functioning level, capacity allowing renewal of all functions and limitation of long-term breaks 5. Constant education process, requirement to use circular planning systems based on the Deming Cycle, allowing to introduce new solutions in case of changes in surrounding environment. From here, the creation of circular cities is only a step away.

CHOSEN CASE STUDIES: ON THE VERGE OF CIRCULAR

"Green buildings", are a system approach supporting city development and widely promoted by architects and other designers (Yudelson, 2008 and Kibert, 2013). This approach also includes efficient energy use, implementation of sustainable building materials, reduction of waste, but most of all it deals with the design and modernisation of existing buildings and whole complexes. Beginning of the 21st Century was the milestone for this new approach towards the preservation of natural environment. Some of the buildings were actually designed on the verge of circular economy ideas, even if the designers did not as yet have them in mind.

Case of 60L

Such a case is Australian 60 Leicester Street opened in 2002 as one of the first green buildings in this country also using a unique green lease management contract where the tenants have a responsibility to efficiently operate within their lease space and enhance environmental benefits. This was a joint investment of Green Building Partnership (GBP) formed by Surrowee Pty Ltd and Green Projects Pty Ltd, Australian Conservation Foundation (ACF) and University of Melbourne, Department of Architecture Building and Planning. It was designed by an interdisciplinary team of Spowers Architects and Advanced Environmental Concepts and included modernization and extension and re-use of a 19th Century warehouse building located just off the downtown area. Design team included efficient energy and environmental solutions already on the concept level, and developed them during later stages of design which included also recycled fit-out components and finishing materials. Environmental engineers prepared initial design solutions including basic form of the building – a long narrow west-east atrium (Figure 2a), daylight wells on the south and north part of the structure. This simple concept was then passed on to the architects in order to solve functional layout and most efficient use choices. Concept phase was designed by an interdisciplinary team of architects, structural engineers and other technical consultants, including geodesic officer, quantity surveyor, acoustic specialist and a landscape architect. Each design phase was subjected to a close analytic scrutiny check, in order to make sure that the designers were actually fulfilling environmental aims provided in the investor's brief. Presented aims included a choice of building and finishing materials and installation systems which were either recycled or could be recycled within a re-use process at the end of the building's life.

As sustainable development aims are very complex, it was decided that this project would follow four basic assumptions including:

- Use of recycled materials – use of existing building materials including analysis of the environmental influence when using only new building materials;
- Efficient energy use and reduction of GHG emissions – analysed during design, construction and purchase and transport of materials in order to achieve the lowest best case emission scenario, efficient use of daylight and natural ventilation;
- Efficient use of water sources and potable water – use of energy and water efficient appliances, on-site collection and filtration of rain water, reduction of chemical substance use, circular use of water sources;
- Human connections creating societies with a mutual aim to lower environmental influence during development and implementation of an Environmental Construction Management Plan; efficient use of buildings through implementation of green tenant leases, a high quality tenant monitoring system with individual tenant meters measuring the level of energy and water used in a agreed to period of time, not exceeding the level of energy and water specified in the green lease contract.

Energy system is supported by a set of photovoltaic roof top panels and used during weekends and outside the energy peak hours (it sustains circa 10% of the building's needs). Another interesting analysis (supported by a World Green Building Council Research from 2018, The Business Case for Health and Wellbeing in Green Building) showed that outside economic benefits, this building actually improved occupant satisfaction and productivity by 5-15% in comparison to other typical office buildings.

Figure 2a. L 60 Building view of the long internal atrium

Figure 2b. Local type of plants used on the green roof garden

Figure 2c. Rainwater tanks placed in the main atrium on the ground level of the building
Source: (photographs: Rynska E.)

Figure 2c shows rainwater tanks located on the building's ground floor in a place which is clearly visible to all visitors and building users. Water is one of the most important cicular elements in Australia which suffers from periodical draughts and high temeperatures. Used solution proved that rainwater can be used also for drinking purposes. Since no direct laws were applicable at that time, facility manager was responsible to check on a daily routine whether existing supply fulfilled potable parameters.

Water is one of the sources with major impact in Australia and some extraordinary solutions were used in L60. This included collection of rain water in a rooftop tank, later filtered and purified to a level where it could be used for drinking purposes. The most difficult issue was that at that time there were not any adequate Australian laws in existence accepting rain water as potable. Investor decided that collected water was analysed on a daily basis to prove that it was fit for drinking purposes. Otherwise this system allowed to reduce city mains water source requirements by 90%, when assuming average standard precipitation level. Building fulfils all user requirements and in extreme situations such as draught, there is always a possibility to use

Figure 3. L60. Ground floor layout showing location of the internal water feature acting as a water tank and other water connection
Source: (based on data sourced from ACF)

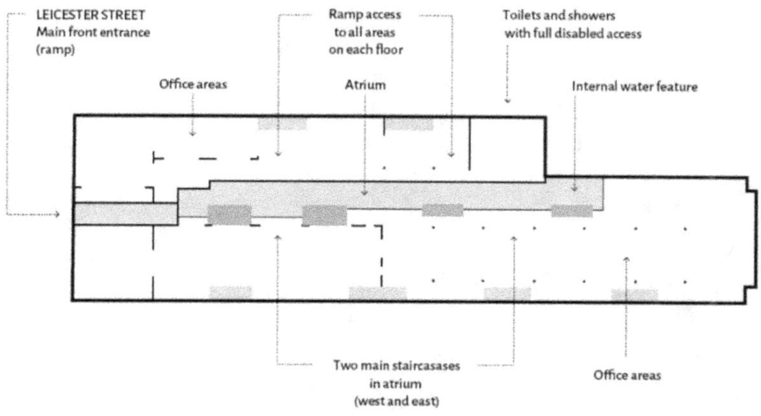

city connection. 60L also fulfils fire prevention standards, and in case of fire outbreak, sprinkler installation is connected to the city water main. Figure 3 presents internal feature – an internal artificial pond – being a part of the rain water retention system. It also enhances solutions proving that the building fulfil public open access requirement including ramps, sanitary areas, lifts and areas width sufficient width for a disabled wheelchair person to make a turn.

Figure 4a presents a water collection system used in L60 Building. Rainwater is captured on the roof (except for small garden area (Figure 2b), and then filtered through the collection system into Ground Floor storage tanks. From there, it is filtered through a subsystem and UV sterilised. In sequence it is supplied to showers, sinks, sanitary basins and kitchens. Filtered water system, when required, may be supplied from Melbourne city mains. Figure 4b, also shows that water used in toilet flushing and waterless urinals is returned to sewage plant and then looped back into sanitary system. Surplus of water which has passed through the roof garden soil is also used as an internal water feature – already mentioned. It includes reeds and water purifying plants to remove nutrients, with excess removed to the city sewer system to form yet another retention loop.

As already argued, sustainable environment does not have a precise definition. It is associated more with particular choices – different construction techniques and materials with reduced primary energy input, lower building waste volume and environmental pollution emissions. These general

Figure 4a. L60 green building – water collection system
Source: (Based on data sourced from ACF)

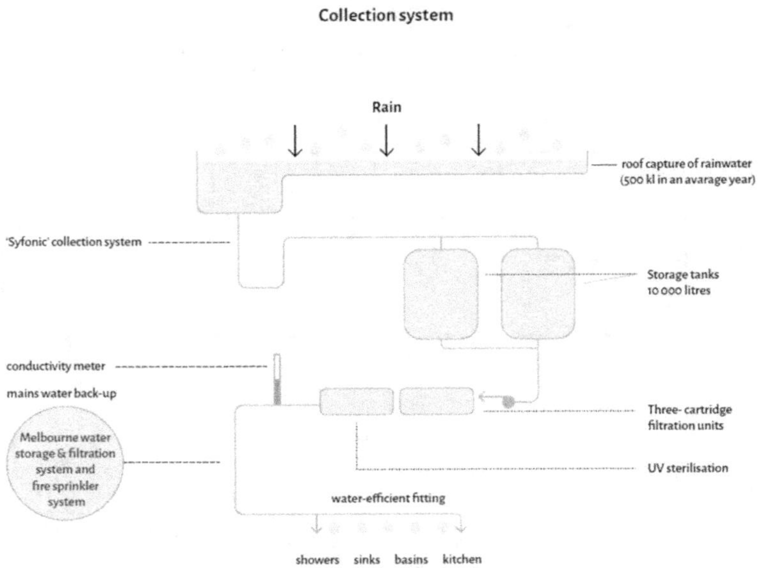

assumptions were actually used by the L60 designers and construction companies and they may be presented in a following hierarchy:

- Reduction of waste production;
- Limitation or lower volume of building materials used, re-use of building materials;
- Use of recycled materials or materials which may be recycled in future after the end of their technical life;
- The least expected and most expensive waste management is simple removal to land-fill – as this choice leads to loss of valuable sources and also takes place.

Australian Conservation Foundation remained building's largest tenant until 2009, when Green Building Partnership the previous owner and operator donated the building to become a ACF's property.

Figure 4b. L60 green building: Water circulation re-use
Source: (based on data sourced from ACF)

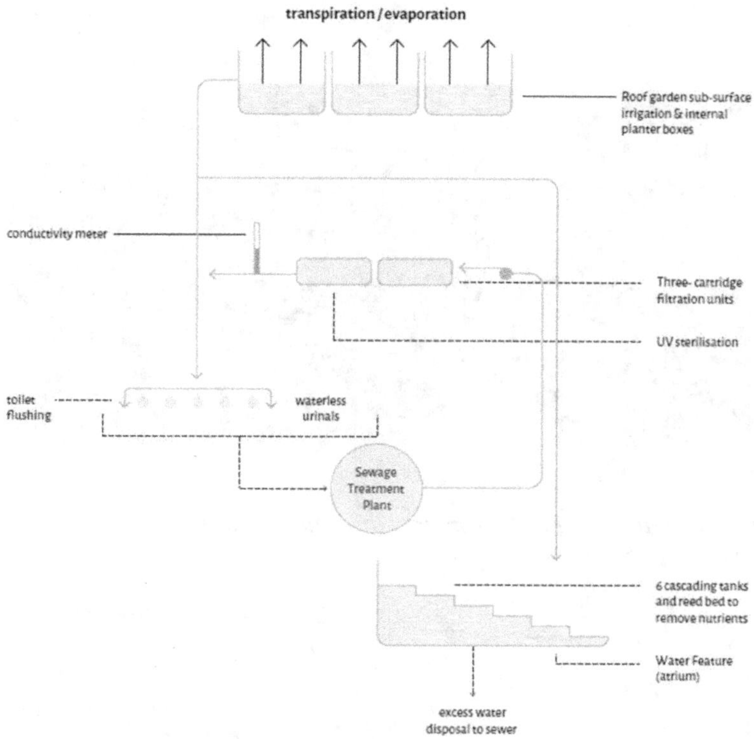

Case of Agenda 51 School

The most common reason for the changing attitude towards design and construction of buildings is a state when user expectations are different from the prevailing standards. In Norway educational reform of the 1990 was a catalyst for such a change. The basic goal was to adjust existing teaching facilities to the educational programme and new teaching methods. Quite soon, teaching methodology merged with sustainable development issues, to form education curricula preparing the society to accept this then new philosophy into everyday life. Hence, part of the changing drive for this pre-circular change was the need for a better understanding of the actual value of resources on which our lives depend on, and school environment proved to be the best choice.

As Norway is one of the leading countries where implementation of sustainable solutions is implemented within many efficient solutions and

Figure 5a. Kvernhuset School view of the main entrance

Figure 5b. Kvernhuset School main entrance hall building
Source: (photographs: Rynska E.)

building standards. From various schools visited by the author, this particular case is best suited to the circular scheme especially where management, energy and water schemes are concerned.

Kvernhuset High School building was designed for the Fredrikstad Commune and constructed in 2002. The on assumptions which were further

Figure 6a. Kvernhuset School view of the front façade with artificial lake in the foreground

Figure 6b. Kvernhuset School view of one of the school's education side wings
Source: (photographs: Rynska E.)

developed on the conditions initially set forth in Agenda 21. The added value in Norwegian school design was the implication that the designers may have a strong influence on creating solutions which enhance the "see" and "touch" teaching methods, as well as "the question method" – why, where and what. Both methods were used in many Norwegian schools, but they appear to be most visible in the presented case.

Outside strong education values and efficient energy solutions including intricate BMS controlling, every smallest internal comfort parameter, this building is on the verge of circularity for following reasons:

- The initial site was located at a certain distance from the town, in a woodland area growing on the soil topping granite boulders; the designers decided that rocks excavated in order to construct foundations were to be used as external wall cladding of the main building (Figure 6a), either as gabion baskets units or simple structure granite walls, whereas seasoned timber from the felled trees was used partly as external finish for the education wings (Figure 6b);
- Public areas were "fitted" with natural - granite boulders used as furniture (Figure 5b);
- Places where the trees used to grow are either "commemorated" as floor-in-fills or in some cases the trunk of trees remain as reminiscent icons reminding, that in order to use this site, a forest had to be cut down;
- Artificial lake in front of the building (Figure 5a) is part of a complicated water purification system with the water reused for the green areas;
- Building is uses alternative energy from a set of deep ground pumps; many of the furniture fit-outs is made from re-used plastics.

Both presented examples present interesting case studies even, if at the time they faced many controversial remarks concerning solutions used during design and construction procedures. More than likely they are not the only ones on the circular verge, but serve as adequate milestones on the road to the circular architecture and urbanization.

CONCLUSION

It should be noted that approach to city development changes together with better understanding of the environmental processes driving our Planet.

Circularity always existed within the widely understood construction industry, but for the past century it was not the main stream option. The designers and investors started to change their attitude at the end of 20th and beginning of the 21st Centuries. This changing approach can be perceived in the case studies presented in this chapter. As already mentioned earlier, presented buildings are not unique of those times and other construction investments developed according in analogous scheme may be found in other locations.

REFERENCES

Australian Conservation Foundation. (n.d.). *60L Green Building*. Retrieved from www.acf.org.au

Brenner, N. (2004). *New State Spaces, urban Governance and the Rescaling of statehood*. Oxford, UK: Oxford University Press. doi:10.1093/acprof:oso/9780199270057.001.0001

Buijze, A. (2013). *Case Study Utrecht Station Area,the Netherlands: How PPPs Restructured a Station, a Shopping Mall and the Law*. Report 4. AISSR programme group Urban Planning. Retrieved from https://www.uu.nl/sites/default/files/rebo-ucwosl-2013-urd-context-report-4_utrecht-station-area.pdf

Centre for Cities. (2013). London: Centre for Cities. Retrieved from http://www.centreforcities.org

Chourabi, H., Nam, T., Walker, S., Gil-Garcia, J. R., Mellouli, S., Nahon, K., . . . Scholl, H. J. (2012). Understanding Smart Cities. An Integrative Framework. In *Proceedings of the Forty-Fifth Annual Hawaii International Conference on System Sciences*. Wailea, HI: IEEE Computer Society.

Esposito, M., Tse, T., & Soufani, K. (2018). Introducing a circular economy: New Thinking with new Managerial and Policy Implications. *Berkeley Haas, California Management Review, 60*(8).

Etkins, P., Simon, S., Deutsch, I., Folke, C., & De Groot, R. (2003). A Framework for the Practical Application of the Concepts of Critical Natural Capital and Strong Sustainability. *Ecological Economics, 44*(2-3), 165-85. Retrieved from http://citeseerx.ist.psu.edu/viewdoc/download?doi=10.1.1.578.4786&rep=rep1&type=pdf

European Green Capital. (2018). Luxembourg: Publications Office of the European Union. Retrieved from http://ec.europa.eu/environment/europeangreencapital/wp-content/uploads/2013/02/Nijmegen_EGCA2018_Brochure_EN.pdf

Ewing, B., Moore, D., Goldfinger, S., Oursler, A., Reed, A., & Wackernagel, M. (2010). *Ecological Footprint Atlas*. Oakland, CA: Global Footprint Network.

Grober, U. (2012). *Sustainability: A Cultural History*. Totness: Green Books.

Hackworth, J. (2007). *The Neoliberal City: Governance, Ideology and Development in American Urbanism*. Ithaca, NY: Cornell University Press.

Hollands, R. G. (2008). Will the Real Smart City Please Stand Up? City: Analysis of Urban Trends. *Culture, Theory, Policy. Action*, *12*(3), 303–320.

Joss, S. (2015). *Sustainable Cities. Governing for Urban Innovation*. Macmillan Education. Palgrave.

Kargon, R. H., & Molella, A.P. (2008). Invented Edens: Techno-Cities of the Twentieth Century. *The American Historical Review, 114*(2), 427–428.

Kennedy, C., Cuddihy, J., & Engel-Yan, J. (2007). The Changing Metabolism of Cities. *Journal of Industrial Ecology*, *11*(2), 18. doi:10.1162/jie.2007.1107

Kennedy, C., Pincetl, S., & Bunje, P. (2011). The study of urban metabolism and its applications to urban planning and design. *Environmental Pollution*, *159*(8-9), 1965–1973. doi:10.1016/j.envpol.2010.10.022 PMID:21084139

Kenworthy, J. R. (2006). The Eco-city: Ten Key Transport and Planning Dimensions for Sustainable City Development. *Environment and Urbanization*, *18*(1), 67–85. doi:10.1177/0956247806063947

Kibert, C. J. (2013). *Sustainable Construction: Green Building Design and Construction*. Hoboken, NJ: John Wiley and Sons.

Lehmann, S. (2010). *The Principles of Green Urbanism: Transforming the City for Sustainability*. London: Earthscan.

Miller, M. (2002). Garden Cities and Suburbs: At Home and Abroad. *Journal of Planning History, 1*(1), 6-28. Retrieved from https://journals.sagepub.com/doi/abs/10.1177/153851320200100102

Newman, P., Beatley, T., & Boyer, H. (2012). Resilient Cities: Responding to Peak Oil and Climate Change. *Journal of Urban Design*, *17*(2).

Oatley, N. (Ed.). (1998). *Cities, Economic Competition and Urban Policy*. London: Paul Chapman Publishing.

Pearce, D. (2002). An Intellectual History of Environmental Economics. *Annual Review of Energy Environments, 27*, 57-81. Retrieved from https://www.cepal.org/ilpes/noticias/paginas/1/35691/ja_histofenvecon.pdf

Pearce, D., & Barbier, E. B. (2000). Blueprint for sustainable economy. *Business Strategy and the Environment, 10*(4).

Pearce, D. W., & Turner, R. K. (1990). *Economics of Natural Resources and the Environment*. New York: Harvester Wheatsheaf. Retrieved from https://www.sciencedirect.com/science/article/pii/0308521X9190051B

Peck, J., & Tickell, A. (1994). Searching for a New institutional Fix. In A. Amin (Ed.), *Post-Fordism: A Reader* (pp. 280–315). Oxford, UK: Blackwell. doi:10.1002/9780470712726.ch9

Pulselli, R. M., Magnoli, G. C., & Tiezzi, E. P. B. (2004). Energy Flows and Sustainable Indicators: The Strategic Environmental Assessment for a Master Plan. *WIT Transactions on Ecology and The Environment, 72*, 3-10.

Rees, W. E. (1992). Ecological Footprints and Appropriated Carrying Capacity: What Urban Economics Leaves Out. *Environment and Urbanization, 4*(2), 121-30. Retrieved from https://journals.sagepub.com/doi/10.1177/095624789200400212

Rees, W. E., & Wackernagel, M. (1996). *Our Ecological Footprint: Reducing Human Impact on the Earth*. Gabriola Island, BC: New Society Publishers.

Register, R. (1987). *Eco-city Berkeley: Building Cities for Healthy Future*. Berkeley, CA: North Atlantic Books.

Rockefeller Foundation. (2013). *Rebound: Building a more Resilient World*. Retrieved from https://assets.rockefellerfoundation.org/app/uploads/20130124192107/Rebound-Building-a-More-Resilient-World-.pdf

Rydin, Y. (2010). *Governing of Sustainable Urban Development*. London: Earthscan.

Smith, R., Simard, C., & Sharpe, A. (2001). *A Proposed Approach to Environment and Sustainable Development Indicators Based on Capital.* Report prepared for The National Round Table on the Environment and the Economy's Environment and Sustainable Development Indicators initiative. Retrieved from http://www.oecd.org

Stern, D. I. (1997). The Capita Theory Approach to Sustainability: A Critical Appraisal. *Journal of Economic Issue, 31*(1), 145-173. Retrieved from https://www.tandfonline.com/doi/abs/10.1080/00213624.1997.11505895

UN-Habitat. (2003). *Sustainable Local Government, Sustainable Development: Involving the Private Sector and NGOs*. Opening Keynote Speech, 4 March. United Nations Human Settlements Programme UN Habitat. Retrieved from http://www.unhabitat.org

UN-Habitat. (2009). *Global Report on Human Settlements 2009: Planning Sustainable Cities*. London Earthscan. Retrieved from https://unhabitat.org/books/global-report-on-human-settlements-2009-planning-sustainable-cities/

UN-Habitat. (2010). *State of the World's Cities 2010-2011: Bridging the Urban Divide – Overview and Key Findings.* Nairobi: United Nations Human Settlements Programme UN Habitat. Retrieved from https://sustainabledevelopment.un.org/index.php?page=view&type=400&nr=1114&menu=35

UN-Habitat. (2011). *Global Report on Human Settlements 2011: Cities and Climate Change: Policy Directions. Abridged Report*. London Earthscan.

UNDESA. (1992). *Earth Summit Agenda 21: The United Nations Programme of Action from Rio*. UN Department of Economic and Social Affairs, Division for Sustainable Development. Retrieved from http://sustainabledevelopment.un.org

United Nations. (2013). *Towards Sustainable Cities – the United Nations.* Retrieved from http://www.un.org/en/development/desa/policy/wess/wess_current/wess2013/Chapter3.pdf

WCED - World Commission on Environment and Development. (1987). *Our Common Future*. Oxford, UK: Oxford University Press. Retrieved from https://sswm.info/sites/default/files/reference_attachments/UN%20WCED%201987%20Brundtland%20Report.pdf

Weisz, H., & Steinberger, J. K. (2010). Reducing Energy and Material Flows in Cities. *Current Opinion in Environmental Sustainability*, *2*(3), 185–192. doi:10.1016/j.cosust.2010.05.010

Wheeler, S. (2000). Planning for Metropolitan Sustainability. *Journal of Planning Education and Research, 20*, 133-145. Retrieved from https://journals.sagepub.com/doi/10.1177/0739456X0002000201

Wheeler, S., & Beatley, T. (Eds.). (2009). *The Sustainable Urban Development Reader* (2nd ed.). London: Routlage.

Yudelson, J. (2008). *The Green Building Revolution.* Island Press.

Zhang, Y., Yang, Z., Fath, B.D., & Li, S. (2010). Ecological Network Analysis of an Urban Energy Metabolic System: Model Development, and a Case Study of Four Chinese Cities. *Ecological Modelling, 221*(16), 1865-79.

Zhou, N., & Williams, C. (2013). *An International Review of Eco-City Theory, Indicators and Case Studies*. Berkeley, CA: Ernest Orlando Lawrence Berkeley National Laboratory. doi:10.2172/1171800

KEY TERMS AND DEFINITIONS

Agenda 21: A non-binding action plan of the United Nations with regard to sustainable development. First introduced in 1992 at Earth Summit Conference in Rio de Janeiro, Brazil.

ENDNOTE

1 European Green Capital Award – is a result of an initiative undertaken 15.05.2006 by 15 European cities (Tallinn, Helsinki, Riga, Vilnius, Berlin, Warsaw, Madrid, Ljubljana, Prague, Vienna, Kiel, Kotka, Dartford, Tartu & Glasgow) and an Association of Estonian Cities, signed in Tallinn. This Treaty established an award for the city leading in the initiatives for establishment of environmentally friendly housing environment. Glasgow).

Chapter 3
Circularity in Modern Cities

ABSTRACT

This chapter is dedicated to information on the macro trends shaping the future of our contemporary urbanisation processes, including environmental support of the city development and management issues, definition of factors that should allow the transformation of the existing cities, and possibilities for future changes and evolutions. Case studies of European cities management will be presented. The main issues discussed will be approach to the circular of energy and water sources, air and building materials, and the rationality and efficiency of their use.

CIRCULAR APPROACH TO CITY MAKING

Circular economy basics appeared as early as in the -70-ties of the 20th Century with the intention to promote a world without waste. Over 40 years later this vision became the main aim for many country governments. Analysis provided by Kircherr, Reike and Hekkert (Kircherr et al, 2017) allowed to formulate a holistic definition according to which circular economy is described as an economic business concept which will take the place of the "material's end of life" and concentrate on limitations, alternative use, recycling and reclaim of materials during production and consuming processes. Hence, the changes will take place on the micro-levels (products, firm and consumers), intermediate levels (eco-industrial parks) and macro-levels (cities, regions, countries and larger regions). All above tasks, lead to further sustainable development and

DOI: 10.4018/978-1-7998-1886-1.ch003

pursue environmental quality, economic prosperity and social quality. These will become an added value for the present and future generations.

Circular economy challenges traditional linear economy in the areas concerning extraction, growth, production and transportation of resources. Traditionally, they are discarded prior to their maximum use value which can be achieved in the processing chain. This model was effective, only when the in-put goods were cheap and easily accessible, presently this approach does not lead towards sustainable development, hence the requirement to "close the loops" and approaches to create an economic sustainable development model. A scheme presenting a transformation sequence of a process leading to circular economy standard is depicted on Figure 1. Circularity is increased through application of triple steps: useful application of materials, extended lifespan of products and parts, as well as smarter products use and manufacture. Above concepts are managed by various tasks – strategies. The first level – useful application of materials – consists of recycle and recover issues, both being presently the best developed areas in construction industry. This also is the lowest level and changes and the next two, are in many cases still out of industrial reach. Nevertheless extended lifespan is can already be traced in areas where reuse, repair, refurbish and repurpose applies. Remanufacture in building industry is still an area which should be part of the research area. Smarter products may also be found on the market, but the application of refuse, rethink process are still to be researched on a scientific level. Whereas application of reduce has strong possibilities in building and the management process should be more efficient both where it comes to production and construction processes.

Basics of the circular economy approach can be found in many theories dating back to the former century, but as stated in the first lines of this chapter, only last decades brought concrete research and conditions for further development. Some of the related concepts presented in detail by Ellen MacArthur Foundation are presented below (www.Ellenmacarthurfoundation.org.):

- Sustainable development is a holistic concept which mainly deals with growth of three areas: economic, environmental and social. It has connections with the circular economy mainly due to its concern towards social responsibility of manufacturing companies described as the threshold to sustainable development;

Figure 1. Transformation of the linear economy into circular
Source: (Circular Economy in Cities 2018.)

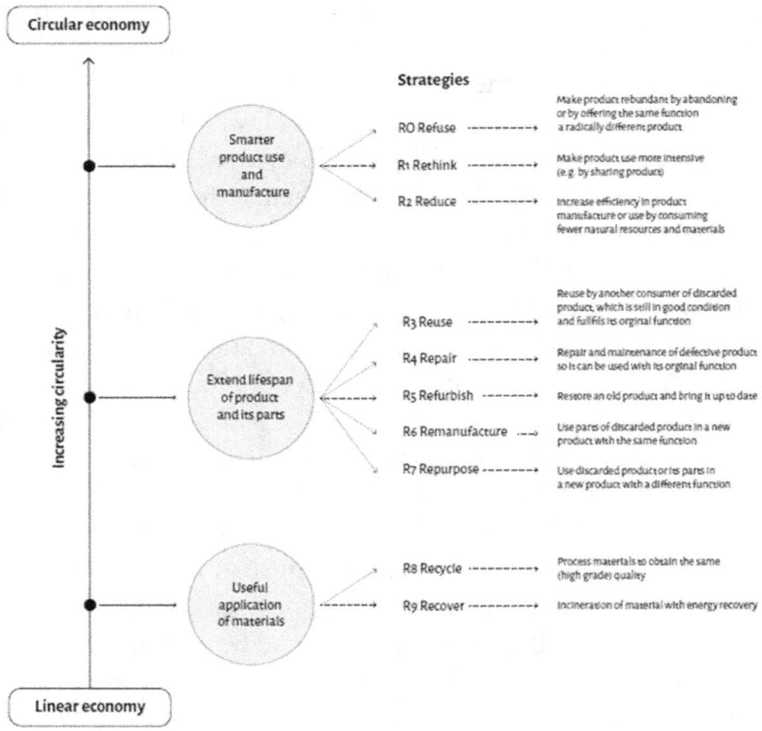

- Green economy promoting economic solutions for environmental issues, to be implemented through international management as initially proposed by the United Nations. Implementation will be provided by the managerial bodies presented by the governments of various countries and the NGO's. This approach has common interests with circular economy, as both use economy in order to achieve sustainable development goals;
- Functional economy, together with the Cradle-to-Cradle theory and industrial economy, is one of the main paths within circular economy. Its basic conditions point out that in order to achieve division between economic growth and the level of resource consumption, new business models have to be implemented. These models should act more as a service or efficiency providers, than services with maintained ownership. Proposed idea may initiate innovative approaches and the act of implementation will become a threshold when the production

and consumption loops will start to close. Functional economy in this case is demonstrated through intelligent use of waste products and realized by innovative businesses;

- Provision of LCT is closely related with acceptance of conditions estimated by Life Cycle Assessment (LCA) and Life Cycle Management (LCM). The main aim is to lower environmental impacts through implementation of ISO standards. This ecologically efficient methodology may be used during manufacturing phases, and during the user phase. Interdisciplinary design and re-design processes aim at the lowest possible impact. This idea contains systematic concept conditions (LCA and LCM) imbedded in the value chain approach, and includes relative estimation of sustainable values (i.e. estimation of negative impacts related to the critical areas found in the analysed process). Whereas circular economy uses above scheme throughout sustainable development – absolute sustainability assessment (ASA);
- Cradle-to-cradle Concept is based on the efficient eco-solutions determined by recycling processes of technical components (i.e. durable object manufactured from plastic and metal) and biological components (i.e. nutritive products). This process will in turn allow for a healthy metabolism of materials used, analogously as in case of industrial symbiosis programs (Kalmykova et al. 2015). Concept, authored by William Mc Donough and Michael Braungart, creates design framework encompassing circular economy issues. It also includes a circular paradigm concerned with the ability to achieve a full recycling process, an outcome of an adequate design and re-design of products with the use of safe and healthy components and LCA input. In this case, every effect assumed by circular economy is optimized. In early 90-ties of 20th Century, the authors implemented their biologic and technical approach under "services" heading;
- Co-ownership practice is a management approach created by Michael Porter and Mark Kramer. The leading aim was to find a platform of understanding between main stream Capitalistic ideas and social requirements. Idea evolves around the need to create values through identification and support given to social expectations. Manufacturing processes of the new products and stable economic markets, redefinition of the value chains and creation of social development clusters by business stakeholders is also part of this concept. The main aim of this new value is creation of connections between evaluation approaches and business strategies aiming at economic gains. Both,

the practice of co-gains and the circular economy concept are based on the condition that they should undergo restructuration. Additionally, certain mechanisms should be implemented in order to sustain further business and consumption development. It should be noted that circular economy actually propagates implementation of much wider changes;

- Industrial ecology idea was formulated within a zone of scientific and implementation mechanisms concentrated on the creation and support of closed loop ecosystems used in industrial processes. The main aim is efficient use of energy and material resources, reduction of polluting waste and by-products through their economically justified transformation into useful end-products. The final aim is to create industrial systems mimicking environmental systems (bio-mimicry). Scope of application is based on existing manufacturing chains between various industrial enterprises producing goods in eco-cycles. This approach is in accordance with the circular economy system conditions where efficient loop use of resources is concerned;
- Extended Producer Responsibility (EPR) is an idea where the polluter is also the tax-payer and includes the requirement to extend the manufacturer's environmental responsibility for the product throughout the product's life. Main conditions concentrate in the management zone starting after consumer phase, but include the possibility to make modifications during the product's whole life cycle. EPR, is a business concept with its application concentrating on a single enterprise. It is the initial application case of the closed loop systematics prepared as-a-suit-for-purpose;
- Eco-design is concerned with the integration of environmental aspects during product's creation process. This approach may be used as a tool allowing for the implementation of LCA effects, and a set of framework conditions – a list of conditions or analytic tools supporting the eco-efficient product during manufacturing process. Eco-design is a tool which supports implementation of sustainable development issues within the product's design phase and is often used simultaneously with LCA.

Initially, environmental design philosophy concentrated on the selection of components which could undergo recycling processes. Presently, it has evolved into a process which includes radical redesign of products and services according to ecologic, economic and social expectations mirroring sustainable development requirements. Sustainable use of resources and services, as well as development of extensive economic markets supporting products' durability have become important parts of the proposed production cycles. Due to the materials' typologies and flows of individual systems, circular economy may be perceived in two dimensions. The upstream phase – prior to the use phase – is concerned with efficient management of resources, efficient use of components during manufacturing phase, reduction of waste products and maintaining of the lowest possible manufacturing costs. The other dimension is the downstream phase, concerned with maintaining values of the remaining waste and optimizing extraction processes within proposed production framework. A good example might be: less waste volume delivered to the land in-fill sites due to retrieval and recirculation of valuable components in economic processes. Ellen Mc Arthur Foundation has defined three conditions characterizing circular economy (Ellen McArthur Foundation, 2015), being also in counterbalance with Lacy and Rutqvist ideas (Lacy and Rutqvist, 2015):

- Preservation of Values: Concerned with preservation of the highest possible component's value used during manufacturing processes, modelling, and in the finally achieved product. This approach is concerned with modifications in the final economic aims, changes during the manufacturing and modernization processes, repair and re-use of the components, achievement of a longer life cycle and products' value cycles;
- Efficient Use of Resources: Applies to an efficient and lower than presently use of basic resources and includes an optimized process of waste selection, recycling processes, recuperation of energy during incineration and use of alternative energy resources;
- System Efficiency: Concerned with a systematic reduction of waste derived from all components valuable to human civilization (i.e. nutrition, shelter, health, education) during the manufacturing and consuming processes and includes external factors (i.e. use of soil surface, air and water volume and acoustic pollution as well as climatic changes).

Circulation of building materials is dependent on the national, regional or city policy (Barles, 2010) and flexibility of the existing legislation, which may in fact either support or hinder use of the non-standard solutions. Efficient state policy is one of the key factors which enables sustainable flow of resources in city areas (Quian et al. 2013). Materials with low market value (most of the building waste, except for glass, steel and other metals) in most cases are placed in waste dumps i.e. effective selection must be done as a state intervention, including introduction of adequate financial means and mediations with stakeholders involved already in the initial phases of sustainable waste management (Sheu, Chen, 2014). In countries with over 80% of recycling processes which are driving towards closed loop economy i.e. Germany (86.3%) and Holland (98.1%) (Tojo, Fischer, 2011), re-use and effective use of materials is treated as a priority. German and Dutch legislation (Kreislaufwirtschaftsgeset, 2013; Landelijk afvalbeheerplan, 2014), according to EU directives (Directive 2008/98/EC), also point towards sustainable material circulation. They implement the basic legislation rules to be later on developed in EU member countries as closed circular loops dedicated to particular substances (in Germany) and material sequences (in Netherland). Prevention, reuse and recycle processes are preferred instead of incineration or waste storage. These legislation acts regulate the issue of waste during the materials' total life cycle, and the prevention procedures are analysed prior to the product becoming waste i.e. in the design phase or pre-construction in case of existing buildings which will be demolished. Waste legislation acts are supported by such documents as German regulations for the commercial and construction waste or Dutch sector plans (Landelijk afvalbeheerplan, 2014) which define precise standards. The also introduce targets – planned recycling level and minimum standard levels, which define most efficient procedures for each of the selected waste streams. Additionally, sustainable material flows are shaped by national development policies in i.e. Germany (National Raw Material Strategy, 2010) and in Holland (Green Growth Strategy, 2011). These documents prioritise environmental tasks in all development sectors (i.e. green economy). In those countries economic development conditions have to include reduction use of natural resources, efficient use of all materials and a lower impact of human civilisation on surrounding environment. Similar strategies have been introduced on the regional level in Germany in Rhineland-Palatinate (The Circular Economy State of Rhineland-Palatinate, 2008) and in Holland – Limburg. In the first case main impact is placed on the most efficient solution for the circulation of waste and materials in each region and is promoted by the social acceptance of recycling procedures. In the second case, Cradle-to-cradle conditions became the base for the regional policy, and the main aims are preservation of natural resources and non-renewable energy sources. Similar aims have been established in Luxemburg, where circular economy conditions are implemented as a holistic industrial symbiosis also established on the Cradle-to-cradle concept (Schosselet, 2018). Prepared analysis flows and material and energy resources allow the redesign of the technical and organic material flows. This symbiosis is based on a concept where different stakeholders have to cooperate during manufacturing processes. Quality and quantity aims are established individually for each foreseen level, and social actors are also engaged (Schosselet, 2018).

Circular solutions may be also pursued in city policies. British Peterborough will function in accordance with the closed loop economy prior to 2050. Circular economy is also visible in the new strategies of Amsterdam, Rotterdam, Haarlemmermeer and Glasgow (Dhawan, 2018). Mayor of London together with the London Waste Recycling Board have decided that their administration area will include circular economy conditions as priorities (Taylor, 2018). It is therefore visible that similar aims are managed through various alternative routes.

In Paris, circular economy is a vision uniting development processes outside the city jurisdiction. In London, there is a wide promotion of enterprises where low emissions products are manufactured with sustainable use of materials and energy (Fratini et al. 2018). Circular economy can be also perceived in the city revitalisation sector. Closed loop economy conditions are the base for the tasks undertaken in the largest European revitalisation area OPCD Area (Old Oak Park Royal) in London, which is a local planning unit with an area circa 650ha. A Material Flow Analysis (MFA) tool was used during the initial approach to the OPDC Area planning procedures. It included mapping of the material flows in this area (Hammer et al. 2003). Various alternative efficient re-use procedures for local waste were discussed together with the initial outcomes and included in the Local Master Plan. It has been generally accepted that the waste energy strategy and foreseen re-use and recycling procedures are the case best choices (Domenech, Borrion, 2018). In case of the Old Oak Park Royal (2015), buildings are designed and build in technologies which in future will allow easy modification and adaptability to changing user requirements. Volumes are created in accordance with efficient material flows, easy to dismantle and with a possibility to re-use large building elements in other volumes. Described investment includes development of a new social centre for the Western London inhabitants. City flow analysis and closing of the material, substances and energy loops according with circular economy conditions are also the base for the revitalisation processes foreseen for the post-industrial Goudse Haven area in Dutch Gouda City. Urban Metabolism Analysis Tool (UMAT) was used to map the nutrition, bio-waste, material, energy and water flows in this area. Unfortunately, presently analysed flows are short and unconnected, whereas this region is in fact strongly dependent on wide regional connections. Prepared analysis proved that the transformation of this distribution area into a self-dependent zone with manufacture functions, will reduce the losses created by the transport, purchase of energy and environmental taxes.

CASE STUDY: SPANISH BILBAO CITY – A CIRCULAR CITY

Implementation of circular city conditions may be also perceived in the development policy of the Spanish city of Bilbao and the Basque Region, where pioneer implementation solutions have been realised. The Circle City Scan was initiated by a consortium comprising of Innobasque, Circle Economy, Bilbao Ekintza BEAZ. This document is a "road map" identifying possibilities to introduce circular economy procedures and supports preparation of practical and city scaled solutions. It also allows to identify initial tasks, analyses material flows and implementation possibilities including indispensable procedures. Inter-sector analysis also showed most efficient implementation loop procedures, discussed the situation of the restaurant sector, wholesale trade and metallurgy sectors. Six priority development strategies for a circular city were prepared. These are: re-use of nutrition surplus in restaurants, introduction of software innovations to prevent food waste, preparation of a new waste segregation system, use of innovation solutions to optimise wholesale trade, use of 3D techniques in the metal sector and promotion of circular business models. Described strategies are based on seven key elements: preference of alternative renewable resources, preservation and a longer life cycle of existing products, up-date of existing business models, use of waste as resource, optimisation of the city material flows with the use of digital techniques, co-operation between stakeholders and industrial sectors (see Figure 2). Good practises were established for each of analysed areas i.e. for the metallurgy sector – decarbonisation, urban mining (Brunner 2011) with new materials harvested from the city area, re-use of manufacture waste, use of 3D techniques (i.e. to manufacture from destroyed components), effective use of existing resources and search for environmental substitutes. One of the side areas is awareness of circular business models, as without adequate knowledge none of the proposed models can be applied and understood by stakeholders.

Figure 2. Bilbao "road map" leading to circular economy development
Source: Circle Economy, Bilbao Ekintza y BEAZ, (Circle City Scan, Bilbao, 2016)

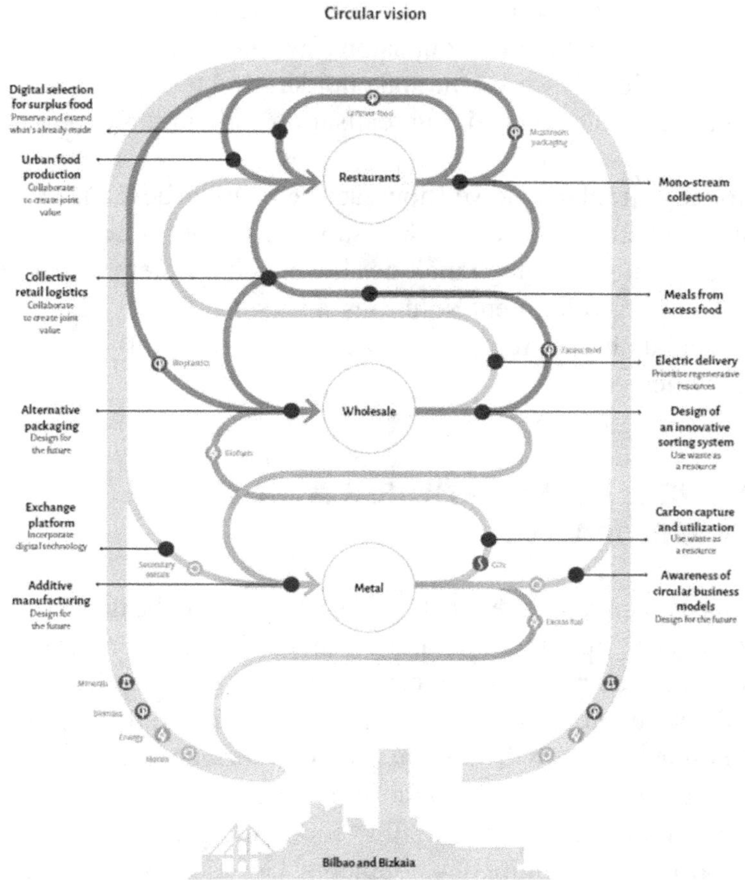

Such areas as digital selection for surplus food and urban food production also complete the approach from the nutritive side of management approach, completing the loops also where food chains can be included.

For the metallurgy sector strategies include: establishment of the software platform (i.e. using Urban Mine Knowledge Data Platform www.urbanmineplatform.eu, European Knowledge Platform on responsible mineral supply chains www.stradeproject.eu), additive manufacture (i.e. use of 3D techniques by Renishaw British enterprise, where bicycle frames are manufactured), re-capture and L60utilisation of carbon (i.e. American bio-technology Algenol www.algenol.com where various carbons are modified into bio-gas), circular business models (i.e. product optimisation as in case of a Belgian WorldAutoSteel www.worldautosteel.org where the weight of the manufactured car was reduced by 270 kg during its entire life cycle, which means reduction of dangerous emissions by 3-4.5 t). In order to establish synergic connections between the stakeholders and tasks, all strategies were collected in a schedule determining realisation and implementation dates for each particular task.

Additionally, the plans showed how each task should be realised by pointing out required stakeholders and indispensable investments as well as possible gains. Circular Bilbao strategy includes lower and efficient use of resources, creation of new development solutions and work places. Loop solutions should also create higher regional autarchy which currently is dependent on external supplies.

CASE STUDY: RHINELAND-PALATINATE REGION, GERMANY

Rising costs of energy sources costs are as much due to the needs and changes currently emerging in our contemporary industrial societies. Such approach is closely connected not just with control of toxic gases and substances, but also with the access to new products and services. It should be indicated that future economic markets will be environmentally driven. German stakeholders have moved their interest zone to those business areas some time ago. This transformation allowed them to double economic gains and create a sector of high importance for future developments. Germany is also perceived as a leader in the production of environmentally friendly energies, and Rhineland-Palatinate has a special status in this context (Ministry of the Environment, 2008). Here, economy and ecology stopped being separate areas. Even a couple of years ago co-operation between those two sectors would have been perceived as unusual. Presently integration of environmental preservation issues in the management standards of many various business stakeholders is considered as a leading solution influencing reduction of environmental pollution and lower costs. Circular economy strategy implemented in Rhineland-Palatinate is more restrictive and more legislatively advanced, than the regular German laws mainly dealing with recycling and waste management, including optimisation of waste streams. One of the important instruments is management of materials flows with emphasis on transparency when dealing with integrated flow of sources and manufacturing processes. Through the consumer's phase up to the user's-end-of-life, and including their influence on the regional GDP. Rhineland-Palatinate Circular Economic Partnership envisages such approach as a base for further development of private local businesses. Cooperation between Ministry of Environment and Ministry of Economic Development corresponds with scientific research conducted by the Institute of Applied Material Flow Management, Birkenfeld Campus, Trier University. This approach leads to the introduction of a new type of co-operation enterprises concerned with efficient use of energy and resources. Representatives of local administrations, universities and stakeholders work together in order to achieve the level of management required for the implementation of circular economy standards. Technical, economic and administrative issues are perceived as thresholds to be crossed when achieving innovation levels. Also many housing neighbourhood zones are active in the implementation process of procedures leading to sustainable development, and in turn loop development. Institute of Applied Material Flow Management is also a Partner in many Project giving scientific support where required, participating in numerous meetings and workshops where the outcomes of scientific research are presented to inhabitants, energy suppliers and business stakeholders. There are both governmental and private projects i.e. bio-energy farm was created off Weilerbach area where Morbach Energy Landscape is located on a former ammunition dump. In 2006, a biogas plant was opened sourcing agricultural waste, and in 2007 – a palette plant – also donating timber waste to the biogas plant. Simultaneously, the biogas waste heat is being re-used as the medium to dry timber elements. Most of the products are sold locally. Hence, this is a good example of a regional circular economy procedure.

Since the very beginning, the main idea included creation of an education-touristic zone, where one of the main areas is teaching the in-comers about the implementation of the alternative energy solutions. Education routes and information classes are part of the program. This process is also followed by private business integration process, with the owners introducing loop solutions into their manufacturing economies. One of the main participants is environmental university campus located in the site of the former soldier hospital in Birkenfeld. University houses over 2000 students who specialise in the environmental sciences – both dealing with engineering as well as legal processes. Institute for Applied Material Flow Management (IfaS), also provides research on building materials, participants and cooperation within the regional technology network. This environmental education is further enhanced through innovative solutions chosen for the construction of various campus buildings. Eco-concept was used in several design phases, starting from reduced building footprint in order to preserve natural soil and local ecosystems, as well as provide better infiltration of rainwater. Efficient use of water is enhanced through re-use of rainwater. Precipitation is collected into tanks, mechanically purified and later used in the toilet flush system, watering of green areas, cleaning purposes, and as a cooling medium. Rainwater, not trapped in the tanks is redirected to retention areas. Roofs are in major part covered with plants reducing water run-off and enhancing local plant diversification. Building materials were chosen firstly for their low internal energy levels and secondly low emissions of dangerous substances. Some of the components came from recycling processes. Heating is provided by co-generic biomass plant (timber and fermenting communal bio-waste). Energy surplus is supplied to the grid and has given this area the status of a Zero-emission University. Alternative energy sources used for heating and electricity purposes are supported by local photovoltaic systems partially integrated within the main campus building's glass façade. The level of required electric light energy has been lowered due to the integration of the daylight and artificial light systems. Several systems allowing for an efficient supply of fresh air has been used in the newest building. Heat recuperation unit and absorber allow use of heat gains for initial preparation of incoming air. Additional transparent insulation layer is placed in front of some of the walls already with high heat inertia, and supports transformation of the radiant energy into heating. These units provide circa 30MWh of thermal energy supporting heating system during Winter or cooling during Summer. All systems are BMS controlled. Alternative options also include sustainable use of water sources, as additional to already implemented rain water system, and separation of used water into grey, brown and yellow streams – in order to recirculate all nutritive substances. Remaining run-off water is directed into local swamp areas. Project is sponsored by external businesses working together on innovative ideas including commercial parks – Ökomparks. Beside co-generative biomass plant and bio-fermentation unit, Birkenfeld Ökompark also houses seven other firms dealing with software, geothermal technologies, controlling and monitoring systems, biometrics, water purification, waste and building installation systems. Nearby Bumholder Ökompark (UCB) deals with recycling and electronic unit dismantling, as well as conducts research on environmental standards. Both Ökomparks co-operate in various scientific areas.

Analogously as in other German regions, high attention is placed on waste selection. Rhineland-Palatinate Hinkel Netzwerk International is an enterprise which deals with implementation of those solutions in Rhineland-Palatinate, starting from collection procedures, transport and then recycling and waste disposal. Project is still being in the implemented, but the aim is to achieve efficient flow of materials. Other solutions already allowed to lower waste volume. Paper recycling is a very important waste area, it appears to be one of the most valuable circular economy implementations, where energy efficient use of waste heat from plants and equipment, timber pulp and water is used simultaneously during the local purification process. Analogous cycle is used for glass, where recirculation of components allows for a more efficient energy use. These products are labelled with the "Blue Angel" sign. Plastics also belong to the widely used group of products. In circular management various insulation foams or new products are provided during various waste manufacture cycles. Such procedures show the main changes visible in the industry and point out alternative possibilities for circular use.

Additionally, most of the Rhineland-Palatinate inhabitants are part of the water recirculation loop, which is an additional source of potable water. Main purifying plant is located in the Mombach area, where the methane waste products are used during other manufacturing processes. Alternative purification solutions include local plants located in swamp areas, where after treatment, water is relocated to the natural basins. Fehmel purification plant is a good example, where artificial swamp purifies water required to clean vegetables contaminated with artificial fertilisers. Regained fertiliser components are re-used in agriculture, remaining waste is used in the cement production. Sustainable use of environmentally friendly energy sources, including solar energy, is yet another area of interest.

Analysis and realisation of numerous prototype buildings awards data for further innovative implementation. In 2001, LUWOGE (a developer branch of the BASF group) with a seat in Ludwigshafen retrofitted a building, constructed in the 50-ties of the 20th Century, into a 3 litre house. This means that the annual heating energy required for $1m^2$ of housing is below 3 litres of fuel oil. Besides additional insulation and triple glassed windows, heat recuperation from air was also used. Research based on the outcomes allowed for construction of a "zero litre" house. This idea also included management and economic simulations showing how the user or owner will be able to implement solutions without a high rent raise. Efficient energy use was achieved through adequate technical solutions where both classical and innovative solutions were used. Required residual energy is supplied with sun collectors mounted on the building's south façade and photovoltaic roof placed panels. Energy savings are refinanced, as this proposition eliminates water heating costs and lowers building's operation costs. The most important issue is energy optimisation which means 20-30% of heating gains in comparison to a typical housing development. Better efficiency may be also achieved when using alternative energy sources and recycled building materials. Circular solutions also include rain water collected in retention basins and then filtered through soil into natural watercourses and natural landscape parks intensifying commercial, touristic and environmental interests. Beside environmental preservation, such areas may be also used for educational purposes, and prove that sustainable agricultural development could also form a part of the circular economy to be promoted on the European level.

CASE STUDY: AMSTERDAM CITY, THE NETHERLANDS

One the circular economy leaders is the city of Amsterdam, with its Circle Economy vision (Circle Economy, TNO and Fabric, 2016), (Circular Scan Amsterdam, 2016) . City government introduced seven main policy conditions supporting introduction of circular economy issues. These assumptions form a certain vision and also form criteria for a "road map" leading to self-dependent functioning of the city areas. They may be defined as follows:

- **Closed Loops**: Where all materials are part of an infinite technical or biological cycle;
- **Lowered Emissions**: All energy is sourced from renewable sources;
- **Generating Values**: Sources are used to generate financial and other values;
- **Modular Design**: Modular flexible design of products and production chains allowing for a higher adaptation of systems;
- **Innovative Business Models**: New business models for production, distribution and consumption will enable to move the point of importance from ownership to services;
- **Regional Reversed Logistic**: Logistic systems are transformed into regional services with ability to pursue secondary logistic processes;
- **Higher Standard of Environmental Systems**: Human tasks have a positive input into the ecosystems, ecosystem services and reconstruction of natural capital.

Part of the city scan was development as a vision for a circular construction chain. This vision was partly based on interviews with experts and stakeholders. The researchers formulated four strategies which can be used for the city and the whole region, these are presented below (Circular Scan Amsterdam, 2016):

- Smart design with emphasis on a modular and flexible approach, whereby buildings can be updated to new users and other applications and still fulfil current safety requirements and a more durable and functional housing which can be adapted to changing lifestyles and requirements. It is also possible to use 3D technologies as this choice of technology may in near future become more economic (EMF, 2015). A lot of attention is paid to bio-based materials. Over 3 million tonnes of biomass and organic residual waste is released from agricultural areas in the Amsterdam region, from which significant amounts of bio-composite materials can be produced. For example the municipality of Almere has already commenced projects involving bio-waste, which is used to generate bio-composite for the building sector. (Circular Scan Amsterdam, 2016). Nevertheless, it can be clearly seen that in order to move forward there is a need to modify building codes and provide experimental construction areas where proposed solutions will be used in a one to one scale.
- Dismantling and separation is far more efficient when separation of building waste streams is already in place. Destruction currently seems the cheapest option with a cost of only € 20 to € 30 per square metre (Circle Economy, 2016). It is necessary to include decommissioning into the process during early design of buildings. In the circular construction sector, the entire lifespan of a building is taken into account. The costs and benefits of a longer life span are divided between cooperating partners. The cost and time dedicated for each partner are monitored during design, construction, finance, maintenance, operation and demolishment phases. Contractual clauses include conditions for design, building, financing, maintenance and end of building's life and demolition (Netherlands Court of audit, 2013), and formulate requirements to maintain the components and re-use of materials. Separation of construction and demolition waste, allows retrieval of the materials as high-value products without cross-contamination.
- High value recycling applied to a process where building - and waste materials can be reworked into new products. It is foreseen that a special installation will be built in Amsterdam to allow for this process with

a view that it will be accessible for different companies allowing to process and recycle varying streams of construction waste. There is also a set of stakeholders who are researching the possibility of producing CO_2 clean materials. A new project "The Street of the Future'' aims at the solutions to retrieve more materials from street furniture by introducing certain procurement criteria i.e. new products that consist of 70% recycled or reused concrete (The Street of the Future). As many other cities, Amsterdam has an excess of empty buildings which have a large share in the material and energy costs. Major renovation and redevelopment projects, prove that high-value renovation of existing housing can form an effective business case (Kristaps et al. 2016).

- Marketplace and resource bank means that each building can be seen as a material bank of valuable materials. However, presently there is a gap between the demand and supply – dismantling and recycling options. Therefore, a proposition points to the preparation of an online marketplace, supply and demand of building materials for local construction projects can be aligned (by means of GIS data (Zhou, 2014)). Information about the building, management system, building passports, and quality and quantity of materials used in a specific sites can be documented and later made accessible. Also there is a requirement for an advanced collection system and intelligent logistics, which would make the exchange of building materials easier. One of the challenges is storage, in case of Amsterdam the port is proposed as being the central accessible location.

This Amsterdam vision has to face many barriers and obstacles. These implies not only new laws and regulations, but also a change in business culture and much better information channels.

When it comes to a circular management and choice of possible tools, legal acts and management systems implemented on the state level have high impact on the circular economic development. They influence organisation of relations with local businesses. Industrial symbiosis requires dependency between stakeholders and creates chain solutions, where waste from one of the stakeholders may become a resource for another one (Chertow, Ehrenfeld, 2012). Such systems are often based on the individual organisation methods, and their cooperation is motivated through economic and environmental gains (Ashton, 2008). Natural development of chain connections between stakeholders takes long periods of time; i.e. it took over 25 years to create one of the first systems in Kalunborg, Denmark (Velenturf, Ehrenfeld,

Figure 3. Vision of a circular construction chain
Source: (Circular Scan Amsterdam, 2016)

2012. Hence, there is a requirement for initial intervention on the state level (Chertow, Ehrenfeld, 2012), enabling quicker establishment of initial symbiotic conditions (Lewis et al, 2014). Cooperation between service providers, manufacturers and users is based on the exchange processes of resources, and may be stimulated through legal acts, environmental tax system (Patricio et al, 2014) or acceptance of lower waste volumes (Coelho, De Brito, 2013). Symbiotic systems are also created as an effect of extended producer responsibility forming circular economic waste solutions. This law requires the manufacturer to provide financial and managerial procedures for the future waste volume, including disposal. Approach has a high impact on the sustainable recirculation of waste (Zaman, 2014) and modifies connections between stakeholders.

In near future extended producer responsibility will also include building material manufacturers and possibly also building waste. New services will include processes where materials will be lend to the users, and then redeveloped after the end of the life cycle (Addis, 2006). Business based on waste recycling may also be developed as a free access contract signed between small and medium stakeholders with the aim to achieve a set of environmental aims to receive financial grant for an ecology friendly investment i.e. reduction of CO_2 emissions. Donation systems for innovative types of waste collection, recycling and recovery processes, similar to the one existing in Netherlands since 2001, are also helpful. Promotion of environmental design and management through state regulation is very important and efficient i.e. green public procurement (GPP) – a voluntary tool enabling a public framework process where state owned enterprises seek products, services and building works with environmental LCA is lower than in identical cases and procedures ordered without GPP (UZP, 2013).

STANDARDS AND TOOLS CURRENTLY AVAILABLE FOR CIRCULAR APPROACH

Contemporary EU Directives covering environmental issues deal mostly with energy efficiency. Regulations concerning natural resources use standards or reclaiming volumes of materials in construction industries are still not existent. Hence, there is a need to diversify EU Directives with the issues concerning circular flows in construction processes (Boeve, Backes, 2018). Efficient energy requirements must be merged with efficient use of natural resources in order to establish total environmental impact of building processes. Both circulation flows coefficients and particular aims should be established on a continental level, in order to provide a better control over construction sector (i.e. contemporary European standards concerning dismantling of structures are voluntary). Introduction of solutions fulfilling circular economy expectations will require new legal regulations preventing low efficiency solutions, incorrect resource evaluation and inadequate allocation (Tumbull, Tipuric, 2018). There is also a need to introduce more sustainable product management systems. Existing standards regulate manufacturing processes and do not include neither goods, nor their LCA. Environmental issues should be integrated in every phase of the products life and legally regulated. LCA could be the initial factor used in those standards (Malcolm, 2018).

Adequate management systems are also required when introducing closed loop economy. Clear ideas, adequate vision and reputation as well as co-operation of various stakeholders is indispensable (Hopkinson et al. 2018), (Alexander et al. 2018). Following that, open business systems promoting co-operation and open innovation should become an obvious support sector (Chistow, 2018). Such solutions are already under implementation in Rabobank, British Water, H7M or ING. Creation of closed loops and sustainable value Supply Chains requires participation of various stakeholders (from manufacturers to users) throughout the production, use and recovery processes using different activities (Blomsma et al. 2018). Many ideas are already being implemented in Denmark in CIRCit project www.circitnord.com where research includes creation of a method for establishing Supply Chains value, where the main function idea is cooperation between stakeholders (Mishra et al. 2018). Moreover, it is important that the models established should include both professional and legal requirements (Matti et al. 2018).

State development policy shapes the waste management systems with efficiency measures depending on adequate tools promoting retainage processes and sustainable flow of building materials in urbanised areas. Programs are being worked on in Germany and Netherlands: German Environmental Innovation Programme – a large stakeholder fund supporting innovative projects with high potential for environmental reduction; and Dutch Waste as Resource Programme – a state program stimulating sustainable consumption and development, as re-use of materials mainly through education and legislation means. Those countries are also seats of research institutions leading projects on efficient use of materials, Cradle-to-cradle solutions and new technologies i.e. Circular Economy Foundation or C2C Centre Venlo. Beside education and legislation, the key component is also data and accessibility to the harvested building materials and their efficient re-use (Srour et al. 2012). Germany already has MaRess - The Resource Efficiency Network associating various stakeholders who work within the closed loops of building materials. In Netherlands, information on accessible building waste is available in interactive website Oogskart (Superuse Studios from 2007, www.superuse-studios.com). A Harvest Map tool applicable within circa 50km catchment area (depending on the level of traffic, road quality etc.), where potential sources of secondary building materials are mapped (i.e. construction and dismantling sites, manufacturing plants, points of selective waste acceptance etc.). This map may contain accessible materials, dilapidated buildings, potential energy sources, closed sites and unused infrastructure. It also contains information covering location and type of accessible materials, quality and quantity, standard, possible secondary use potential. It supports design procedures when using waste building materials. Similar mock-up maps have been prepared for Rotterdam, Amsterdam, Dordrecht and New York and a free access maps for Brussel region (www.Opalis.be). Integrated Urban Metabolism Analysis Tool (IUMAT) is another tool which maps building material flows which estimates and identifies location sites of materials, energy and nutrition. IUMAT illustrates material in and outflows, enables design of future connections between stakeholders. This program is supported by Building Material Passports (BMP) created by a Dutch architect Thomas Rau. According to his idea buildings represent collection of materials and BMP lists and gives estimation concerning all materials used. An online platform to create and manage material passports of new and existing buildings, and measures circularity and financial value – is Madaster (www.madaster.com/nl).

There are other tools available which are used to measure circularity of cities and buildings. One of them is Urban Circularity Index which with the use of partial coefficients measures city materials flows and resources i.e. uses materials' circular indexes, low emission levels, user health and wellbeing, social participation, infrastructure, mobility levels and cultural diversification (Koehler, 2018). Circularity in buildings may be measured with a Circular Building Assessment (CBA) tool one of the outcomes from a project „Buildings as material banks project BAMB" (financed by Horizon 2020 www.bamb2020.eu). This particular tool measurers dismantling of structural elements or number of components available for re-use. CBA works with BIM system and includes BMP and environmental declaration data. In future this method will enable provision of different scenarios for new circular buildings and define circularity levels for existing buildings (Hobbs et al, 2018). There are also tools for estimating waste management systems, supporting urban planning zones with a closed waste loop – Zero Waste Index (Zaman, 2014). This index is based on 56 coefficients scoping key geographic and administration data, as well as social, cultural, economic, environmental, organisation and political issues.

CITIES AS A KEY TO CIRCULAR ECONOMY

Cities are a key to the transformation from linear into circular economy. They generate more than 80% of GDP which means that they are ideal testing areas for circular economic models. Strong interdependence of business sector, inhabitants and local administration representatives present a starting point for tasks pursued in the living innovation laboratories where complicated economic models are discussed and implemented. It is a rich system both for manufacturers, consumers and agents, with plentiful supply of products and continuous information flow. Due to this, such zones are a adequate for implementation of new practices including changes in reverse logistics systems, collection of data, waste recycling processes and preservation of natural resources, new business models and design of products incorporating circular modes of thinking and procedures. Densely inhabited cities function in an environment with reduced resources, where circular economy effects may be perceived very quickly. Additionally, pragmatism and physical nearness of city areas support implementation of local management much better, than on regional or state levels where bureaucratic structures and rigid decision procedures may halt implementation initiatives. Hence, the cities may be more flexible and may more easily adapt and implement pilot initiatives supporting rapid changes.

According to Ellen Mac Arthur Foundation (Ellen Mac Arthur Foundation, 2015), a circular city includes adequate economy conditions in all of its functions, creating a regenerative and renewable urban system. In such areas the question of waste is eliminated, and any financial assets are maintained as higher use value levels. Wide utilisation of digital technologies is one of the main tools supporting this process. The main task of a circular city is generating wealth and economic resilience for all inhabitants, but there is a strict division between the production values and consumption of the non-renewable resources.

A circular city supports utilisation of system thinking and leads to the achievement of economic, social and environmental gains, while simultaneously searching for solutions supporting a better standard of living. General acknowledgement that the present linear economic model has caused excessive exploitation of resources and increased pollution levels are the turning point to create products manufactured accordingly with sustainable development expectations. Sustainable Development Goals for 2030 (www.sustainabledevelopment.un.org) encourage citizens from all countries around our globe to accept circular economy conditions. Representatives of the largest cities initiate experiments with innovative circular ideas and methods determined by their unique social, cultural, technological and regulation factors. In every case the areas of utmost importance should include:

- **Urbanization Processes**: With the growth of urbanized lands, infrastructure and services initiate a growing negative environmental impact; limited resources have to be used to sustain rapid civilisation growth as well as a growing number of inhabitants;
- **Price and Delivery Risks**: Economic functions of urbanized areas are extremely susceptible to any breaks in resource delivery chains, having further impact on higher prices. Circular conditions may limit this risk through development of industries trading waste as particular required components;
- **Ecosystem Degradation**: Most of the waste volume (solid, liquid, organic and dangerous waste, including that from the construction and demolishment processes) is placed as landfill, which due to its volume and variety forms additional burden for existing ecosystems already devastated by climatic changes; has influence on loss of bio-diversification and causes further degradation and pollution;
- **Environmental Responsibility**: Business and governmental stakeholders are more and more aware of the intensified expectations concerning their responsibilities, as well as reputation hazards where negative environmental impacts are concerned;
- **Consumer Conduct**: Contemporary hyper-consumption means disposal of products before their values have been used, simultaneously there is a growing expectation for new business models where waste volumes may be reduced (i.e. products and services treated as a cooperation system). Experiments with such models in transport and hotel industries appear to draw a lot of positive attention and may be used in other sectors;
- **Development of Technologies:** Digital platforms support use of circular economy solutions on a wider scale through good access to available information, management of materials, logistics, transparency of procedures and responsibility, support adequate placement of innovative circular solutions.

Transformation into circular economy requires rethinking of market conditions, strategies and models which support competition procedures in different sectors and responsible consumption of natural resources. Furthermore, this means a change in the consumption behaviours, generating new working places and limiting requirement for new sources. Therefore implementation of circular economy should enable:

- Creation of new supply chains, where waste from one manufacturing process is used as a source in consecutive one;
- Reclamation of the material resources values in a process which will enable achievement of new values from the same materials;
- Extending utilisation period of products and encouragement procedures to sustain their shares and ownership;
- Better efficiency of the products through shared use.

All shareholders will be able to benefit from the products which have been manufactured according with sustainable development conditions. This can be achieved due to design and manufacture allowing a higher level of possible repair or secondary use, and accompanied with a clear set of information on the sustainable level of sourcing used both for the resources and each component. The inhabitants, beside reduction of their ecologic footprint, will source economic gains from the products durability. If they become shareholders in the initial financial capital, they will be able to generate income derived from the unused, or not completely used products' values. Alternatively, product-as-a-service model, where the clients purchase function or efficient operation instead of becoming product owners, means absence of user servicing, storage, repair or bearing of costs concerned with the end of application phase. For business stakeholders, circular economy gives a possibility to enter into new developments and investment areas such as: collection and reverse logistics, re-marketing, re-use of components and modernisation. Limited use of new resources reduces the possibility of sudden price increase or negative impacts due reduced supplies (natural disasters or sudden political changes). Implementation of circular economy concept promotes a wider evaluation of the supply chains, which may in fact prove to be of a benefit for the resilience of urban spaces. Product-as-a-service model means long term connections between the client and supplier, giving a chance for a durable business relation and a better understanding of the sustainable manufacturing processes (Ellen Mac Arthur Foundation, 2015). For the government representatives, circular economy divides economic development from consumption of the non-renewable resources, allows further development of the city areas simultaneously lowering negative environmental influences. Such development may become a benchmark for the provision of new working places in connection with the high quality recycling processes, appearance of repair and maintenance sector and more efficient logistic processes.

Circular city development is to a certain level associated with the growing number of inhabitants and the need to construct new city infrastructure. Circa 75% of infrastructure required before 2050 has still not be constructed (Global Infrastructure Basel, 2018), whereas required financing for modernisation and construction of new buildings up to 2030 has been estimated as 41 trillion USD (International Resource Panel, 2013). Building materials cover circa 40-50% of the office building carbon footprint (manufacture of cement and steel is equivalent to nearly 80% of energy used during construction process), and this share could be higher, if transport of material and removal of building waste will be included (Circle Economy and Ecofys, 2016). If construction sector will continue to use traditional methods, this could highly influence further devastation of natural environment including air, water and other natural resources, as well as standard of living and economic development. Hence, this could be an incentive for the designers and construction firms to proceed with a more holistic approach when designing, constructing, maintaining and operating buildings, as well as include solutions allowing for further use of building elements and components after the end of their technical life. Therefore, circular city vision must include following areas of interest:

- Urbanized areas designed as modular and flexible, with the use of environmentally friendly building materials which enable higher living standards and reduce quantities of basic used resources. Each building should include the best case building techniques and most efficient functions i.e. co-working office spaces and co-housing areas. Building elements should be maintained and modernized according to the needs, whereas the buildings should generate, not use energies - through use of closed water, nutrition, building materials and energy loops;
- **Energy Systems:** Durable, using alternative energy sources, dispersed, working in efficient energy cycles enabling cost efficiency and lower negative environmental impacts;
- **Urban Mobile Systems**: Accessible, low cost and use efficient; transport systems should include access to public means of transport, including individual cars accessible on demand within flexible "last" kilometre use. Transport units will be fed from electric grid, will be co-shared and automated;
- **Urban Bio- Economy**: Efficient use of nutritive waste is part of the theme; any food particles selected from communal waste, water and sanitary systems to be used as natural composts in urban and

agricultural farming. Urbanized farms will form one of the food sources for the city inhabitants using waste and creating closed loops for further production of nutrition. Energy will be also obtained from rain water and bio-waste, and when sold may become an income for the city areas.

- Production systems should promote creation of local cycles which means higher participation in local production and economic portfolio diversification
- Additional values can be sought in lower transport requirements, and reduced time used for pure communication process; better air and potable water quality can be accepted as additional asset.

Already mentioned urban bio-economy is by far one of the most important areas in the circular economy perspective, as the growing population will expect better quality of life whereas the supporting ecosystems are already overexploited and unstable. The emerging area of bio-economy offers solutions for many challenges, use of renewable biomass and efficient bio-processes being just part of a much wider issue (The Bioeconomy to 2030, 2009). Whereas it is obvious that the major areas of interest will include agricultural and health issues – which are not being part of this book – it will also impact industries, especially if biosensors for real-time monitoring of environmental pollutants and biometrics for identifying people will be introduced as part of the city management program. Moreover, there will be a greater market share allocated to biomaterials such as bioplastics, which appear to be one of major issues when it comes to circular economy. Additionally urban food production can be the case best solution within circular economy transformation which will scope from personal to communal and cover different scales purposes (Beckers, 2018). This should in turn change the outlook of the city area as a place where good quality food for immediate consumption can be produced, which in fact will be cheaper with shorter transport channels. Appropriate urban food sites will have to carefully selected, whereas the waste can be used in the biomass energy sector. According to Steven Beckers (Beckers, 2018) solutions should include the C2C paradigm and include such criteria as: use of healthy upcycle materials, choice of ill-used areas with functions supporting farming (recuperation of waste heat, CO_2 or water), participation in the indoor and outdoor climate quality, public visibility, creation of workplaces as well restoration of local biodiversity.

Reduction of the organic waste volume and capture of the full value of bio-waste streams should be one of the key solutions in the transition towards a circular economy. The biggest potential is locked in the production sector with the efficient use of by-product and waste management. The principle is called profit-driven cascade utilization or bio refining, a method for increasing the value of side streams (Linear to circular, 2018). In Denmark dairy producers use the by-products from cheese production to make high-protein products for the pharma industry and potato processing industries transform potato fibres into a protein-rich food additive. Additionally, final biodegradable waste from these industries is used in agriculture. Applied analysis proved that application of cascading bio-refining within the food and beverage industry on the state level may yield a yearly net value of EUR 300-500 million in Denmark (Linear to circular, 2018). Moreover, Danish Environmental Protection Agency has estimated that 56% of the food waste generated by Danish households may be avoided. In view of this, wide information and education consumer programs should become one of the priorities (Linear to circular, 2018).

Some cities are already introducing new solutions and management chains. One of those cities is Copenhagen where water and energy appear to be the major issues. For example, the Dutch government concentrates on support of urban developments with integrated water strategies. For many

Figure 4a. Kastrup Fort Haven plan of the site

Figure 4b. Kastrup Fort Haven, urban revitalisation development of a former haven into in-city recreation area

Figure 4c. View of the swimming facility at Kastrup Fort Haven
Source: (photograph: Rynska E.)

years swimming in the local harbour was impossible due to high pollution level from sewage overflows, but in 2002 "Islands Brygge" was opened to be followed with other waterfront solutions (Urban innovation…, 2016), a research study prepared by the University of Copenhagen proved that the property value increases by 10%, when there is a park or an urban nature point within a walking distance from the place of living. There are a variety of tools and methods available to integrate water solutions into the urban planning. The most important benefit is less load on wastewater treatment and reduction of sewer overflows. Infiltration of rainwater also increases the groundwater regeneration this last being a valuable city resource.

CONCLUSION

Circular economy challenges traditional linear economy in the areas concerning extraction, growth, production and transportation of resources. In many industrial areas the changes will take place on the micro-levels (products, firm and consumers), intermediate levels (eco-industrial parks) and macro-levels (cities, regions, countries and larger regions). All above tasks, lead to further sustainable development and pursue environmental quality, economic prosperity and social quality. These will become an added value for the present and future generations. Yet, this approach is extremely difficult as it requires transparency as to the standard of the products and components used and co-operation between various stakeholders presenting various types of businesses.

REFERENCES

Addis, B. (2006). *Building with reclaimed Components and Materials.* Earthscan; doi:10.4324/0781849770637

Alexander, A., & Brennan, G. (2018). Linking the Circular Economy field with value modelling in2 the Operations Management, Supply Chain management and Industrial Ecology Literature. In Circular Economy disruptions, past, present and future, International Symposium Abstracts 2018. Ellen McArthur Foundation.

Ashton, W. (2008). Understanding the Organization of Industrial Ecosystems. *Journal of Industrial Ecology*, *12*(1), 34–51. doi:10.1111/j.1530-9290.2008.00002.x

Barles, S. (2010). Society, energy and materials: The contribution of urban metabolism studies to sustainable urban development issues. *Journal of Environmental Planning and Management*, *53*(4), 439–455. doi:10.1080/09640561003703772

Basel, G. I. (n.d.). Guidance Checklists: Preparation of Sustainable and Resilient Infrastructure Projects. Retrieved from http://www.gib-foundation.org/content/uploads/2019/02/Guidance_Checklists_for_Sustainable_Infrastructure_Development_and_Finance_2018.pdf

Beckers, S. (2018). Urban food Production. In P. Luscuere (Ed.), *Circulariteit. Op weg naar 2050?* Tu Delft Open.

Blomsma, F., & Pigosso, D. (2018). Developing Circular Value Chains – Towards a Systematised Approach for Developing Collaboration and Co-creation Efforts. In Circular Economy disruptions, past, present and future, International Symposium Abstracts 2018. Ellen McArthur Foundation

Boeve, M., & Backes, C. (2018). In which ways can law support the transition to a circular construction of buildings? In Circular Economy disruptions, past, present and future, International Symposium Abstracts 2018. Ellen McArthur Foundation

Brunner, P. H. (2011). Urban Mining A Contribution to Reindustrializing the City. *Journal of Industrial Ecology*, *15*(3), 339–349.

Chertow, M., & Ehrenfeld, J. (2012). Organizing Self-Organizing Systems. *Journal of Industrial Ecology*, *16*(1), 13–27. doi:10.1111/j.1530-9290.2011.00450.x

Chistow, V. (2018). Performance assessment of open innovation platforms in circular economy. In Circular Economy disruptions, past, present and future, International Symposium Abstracts 2018. Ellen McArthur Foundation.

Circle City Scan. (n.d.). Bilbao. Circle Economy. Retrieved from https://www.circle-economy.com/how-the-post-industrial-city-of-bilbao-can-accelerate-circular-innovation/#.XTLweOgzY2w

Circle Economy. (n.d.). Developing a Road Map for the first circular City: Amsterdam. TNO and Fabric. Retrieved from https://www.circle-economy.com/case/developing-a-roadmap-for-the-first-circular-city-amsterdam/#.XTMKBOgzY2w

Circle Economy and Environmental Priorities for Business. (2016). World Business Council for Sustainable Development. Ecofys. Retrieved from https://www.wbcsd.org/Programs/Circular-Economy/Factor-10/Resources/Circular-economy-and-environmental-priorities-for-business

Circular Scan Amsterdam. (2016). *A vision and action agenda for the city and metropolitan area*. Amsterdam: Circle Economy.

Coelho, A., & De Brito, J. (2013). Conventional demolition versus deconstruction techniques in managing construction and demolition waste (CDW). In *Handbook of Recycled Concrete and Demolition Waste*. Woodhead Publishing; doi:10.1533/9780857096906.2.141

Dhawan, P. (2018). Circular cities of the 21st century. How are they defined and where are the best practices? In Circular Economy disruptions, past, present and future, International Symposium Abstracts 2018. Ellen McArthur Foundation.

Domenech, T., & Borrion, A. (2018). Application of CE principles to urban regeneration: The OPDC Area London. In Circular Economy disruptions, past, present and future, International Symposium Abstracts 2018. Ellen McArthur Foundation.

Ellen McArthur. (2015). Towards a circular Economy Business Rationale for an Accelerated Transition. Retrieved from https://www.ellenmacarthurfoundation.org/assets/downloads/PL-Towards-a-Circular-Economy-Business-Rationale-for-an-Accelerated-Transition-v.1.5.1.pdf

Fratini, C., Jorgensen, S., & Jorgensen, M. (2018). Exploring the epistemic politics of circular economy: a research agenda with relevance for the governance of sustainable urban transformation. In Circular Economy disruptions, past, present and future, International Symposium Abstracts 2018. Ellen McArthur Foundation. Retrieved from https://www.ellenmacarthurfoundation.org/assets/downloads/Circular-Economy-Symposium-Extracts-June-2018.pdf

Green Growth Strategy. (2011). Letter from the State Secretary for Infrastructure & the Minister for European Affairs & International Cooperation, the Minister of Economic. Affairs, Agriculture & Innovation and the Minister of the Interior & Kingdom Relations to the House of Representatives on the Sustainability Agenda. Retrieved from www.oecd.org

Hammer, M., Giljum, S., & Hinterberger, F. (2003). *Material Flow Analysis of the City of Hamburg, "Quo vadis MFA? Material Flow Analysis – Where do we go?* (p. 2). Wuppertal: Issues, Trends and Perspectives of Research for Sustainable Resource Use.

Hobbs, G., Lowers, F., Balson, K., Fametani, M., Adams, K., & Bourke, K. (2018). The development of a circular building assessment methodology encompassing environmental, economic and social parameters. In Circular Economy disruptions, past, present and future, International Symposium Abstracts 2018. Ellen McArthur Foundation. Retrieved from https://www.ellenmacarthurfoundation.org/assets/downloads/Circular-Economy-Symposium-Extracts-June-2018.pdf

Hopkinson, P., & Harvey, W. (2018). Reputation, leadership and circular economy, Circular Economy disruptions, past, present and future. In International Symposium Abstracts 2018. Ellen McArthur Foundation. Retrieved from https://www.ellenmacarthurfoundation.org/assets/downloads/Circular-Economy-Symposium-Extracts-June-2018.pdf

Kalmykova, Y., & Rosado, L. (2015). Urban Metabolism as Framework for Circular Economy Design for Cities. Proceedings of the World Resources Forum 2015. Retrieved from http://publications.lib.chalmers.se/publication/232085-urban-metabolism-as-framework-for-circular-economy-design-for-cities

Kirchherr, J., Piscicellia, L., Boura, R., Kostense-Smit, E., Muller, J., Huibrechtse-Truijensb, A., & Hekkert, M. (2018). Barriers to the Circular Economy: Evidence From the European Union (EU). *Ecological Economics, 150*, 264–272. doi:10.1016/j.ecolecon.2018.04.028

Koehler, J. (2018). A suggestion for an Urban circularity Index- How can existing findings, concepts, frameworks and standards be combined to serve as indication and alignment for circularity in cities? In Circular Economy disruptions, past, present and future, International Symposium Abstracts 2018. Ellen McArthur Foundation. Retrieved from https://www.ellenmacarthurfoundation.org/assets/downloads/Circular-Economy-Symposium-Extracts-June-2018.pdf

Kreislaufwirtschaftsgesetz vom 24. Februar 2012 (BGBl. I S. 212), das zuletzt durch § 44 Absatz 4 des Gesetzes vom 22. Mai 2013 (BGBl. I S. 1324) geändert worden ist. (n.d.). Retrieved from https://www.bgbl.de/xaver/bgbl/start.xav?startbk=Bundesanzeiger_BGBl&bk=Bundesanzeiger_BGBl&start=//*%255B@attr_id=%2527bgbl112s0212.pdf%2527%255D#__bgbl__%2F%2F*%5B%40attr_id%3D%27bgbl112s0212.pdf%27%5D__1563630669944

Kristaps, K., Blumberga, A., Blumberga, D., Zogla, G., Kamenders, A., & Kamendere, E. (2016). Pre-assessment method for historic building stock renovation evaluation. *Energy Procedia, 113*. Retrieved from https://www.sciencedirect.com/science/article/pii/S1876610217321276

Lacy, P., & Rutqvist, J. (2015). *Waste to Wealth. The Circular Economy Advantage*. New York, NY: Palgrave Macmillan.

Landelijk afvalbeheerplan 2009-2021. Bijlage 6; Invulling beleidskadervoor specifiekeafvalstoffen (sectorplannen). Ministerie van Infrastructuur en Milieu. (2014). Retrieved from https://www.google.com/search?q=-+Landelijk+afvalbeheerplan+2009-2021.+Bijlage+6%3B+Invulling+beleidskadervoor+specifiekeafvalstoffen+(sectorplannen).+Ministerie+van+Infrastructuur+en+Milieu+(2014)+Den+Haag%2C+3.12.2014&oq=-+Landelijk+afvalbeheerplan+2009-2021.+Bijlage+6%3B+Invulling+beleidskadervoor+specifiekeafvalstoffen+(sectorplannen).+Ministerie+van+Infrastructuur+en+Milieu+(2014)+Den+Haag%2C+3.12.2014&aqs=chrome.69i57.1035j0j9&sourceid=chrome&ie=UTF-8

Lewis, K. V., Cassells, S., & Roxas, H. (2014). SMEs and the Potential for A Collaborative Path to Environmental Responsibility. *Business Strategy and the Environment*. doi:10.1002/bse.1843

Linear to Circular. State of Green. (2018). Experiences from Denmark and New York on closing the loop through partnership and circular business models. Retrieved from https://stateofgreen.com/en/uploads/2018/07/SoG_Magazine_Linear_to_circular_210x297_V05_Web-1.pdf?time=1555490701

Malcolm, R. (2018). Life Cycle Sustainability Assessment as a Legal Tool: Regulating for a Circular Economy. In Circular Economy disruptions, past, present and future, International Symposium Abstracts 2018. Ellen McArthur Foundation.

Matti, C., Howie, C., Fernandez, D., O'Sullivan, T., Corvillo, J., & Stamate, E. (2018). Mapping perspectives on sustainability transitions towards circular economy models from a practitioner's perspective. In Circular Economy disruptions, past, present and future, International Symposium Abstracts 2018. Ellen McArthur Foundation. Retrieved from https://www.ellenmacarthurfoundation.org/assets/downloads/Circular-Economy-Symposium-Extracts-June-2018.pdf

Ministry of the Environment, Forests and Consumer Protection Department of Waste Management, Soil Protection, Energy Management, International Environment. (2008). The Circular Economy State Rhineland-Palatinate, Mainz. Retrieved from https://www.stoffstrom.org/fileadmin/userdaten/dokumente/Veroeffentlichungen/Kreislaufwirtschaft_RLP-UK_web.pdf

Mishra, J., Uthayasankar, S., & Lim, C. (2018). Collaboration: An enabler for Circular Supply Chain. In Circular Economy disruptions, past, present and future, International Symposium Abstracts 2018. Ellen McArthur Foundation. Retrieved from https://www.ellenmacarthurfoundation.org/assets/downloads/Circular-Economy-Symposium-Extracts-June-2018.pdf

National Raw Material Strategy. (2010). Securing of a sustainable supply with non-energetic minerals in Germany. Retrieved from https://ec.europa.eu/growth/tools-databases/eip-raw-materials/en/system/files/ged/43%20raw-materials-strategy.pdf

Patrício, J., Costa, I., & Niza, S. (2014). Urban material cycle closing – assessment of industrial waste management in Lisbon region. *Journal of Cleaner Production*, (0). doi:10.1016/j.jclepro.2014.08.069

Quian, S., Zuo, J., Huang, R., Huang, J., & Pullen, S. (2013). Identifying the critical factors for green construction, an empirical study in China. *Habitat International*, *40*, 1–8. Retrieved from https://www.sciencedirect.com/science/article/pii/S0197397513000106?via%3Dihub

Schosselet, P., Shoreder, J., & Mulhall, D. (2018). Implementing Circular Economy Principles in Economic Activity Areas in Luxembourg. In Circular Economy disruptions, past, present and future, International Symposium Abstracts 2018. Ellen McArthur Foundation. Retrieved from https://www.ellenmacarthurfoundation.org/assets/downloads/Circular-Economy-Symposium-Extracts-June-2018.pdf

Sheu, J. B., & Chen, Y. J. (2014). Transportation and economies of scale in recycling low-value materials. *Transportation Research Part B: Methodological*, *65*, 65–76. Retrieved from https://ideas.repec.org/a/eee/transb/v65y2014icp65-76.html

Srour, I., Chong, W. K., & Zhang, F. (2012). Sustainable recycling approach: An understanding of designers' and contractors' recycling responsibilities throughout the life cycle of buildings in two US cities. *Sustainable Development*, *20*(5), 350–360. Retrieved from https://www.researchgate.net/publication/247965642_Sustainable_recycling_approach_An_understanding_of_designers'_and_contractors'_recycling_responsibilities_throughout_the_life_cycle_of_buildings_in_two_US_cities

Taylor, J. (2018). London – An inclusive Circular City? Circular Economy disruptions, past, present and future. In International Symposium Abstracts 2018. Ellen McArthur Foundation. Retrieved from https://www.ellenmacarthurfoundation.org/assets/downloads/Circular-Economy-Symposium-Extracts-June-2018.pdf

The Bioeconomy to 2030. Designing a policy agenda. Main findings and policy conclusions. (2009). OECD International Futures Project, 2009. Retrieved from www.oecd.org/futures/bioeconomy/2030

The Circular Economy State of Rhineland-Palatinate. (2008). Ministry of the Environment, Forests and Consumer Protection, Ministry of the Economy, Commerce, Agriculture and Viticulture, 2008, Druckerei Lindner, Mainz. Retrieved from https://www.stoffstrom.org/fileadmin/userdaten/dokumente/Veroeffentlichungen/Kreislaufwirtschaft_RLP-UK_web.pdf

The street et of the Future. (n.d.). Retrieved from https://resource.wur.nl/en/show/The-street-of-the-future.htm

Tojo, N., & Fischer, C. (2011). Europe as a Recycling Society. European Recycling Policies in relation to the actual. on ETC/SCP Working Paper, 2011. Retrieved from https://portal.research.lu.se/portal/en/publications/europe-as-a-recycling-society-european-recycling-policies-in-relation-to-the-actual(c892ead9-892a-482a-91cb-b426efb8c516).html#Overview

Tumbull, S., & Tipuric, D. (2018). The role of law and regulation in transitioning to a circular economy. In Circular Economy disruptions, past, present and future, International Symposium Abstracts 2018. Ellen McArthur Foundation. Retrieved from https://www.ellenmacarthurfoundation.org/assets/downloads/Circular-Economy-Symposium-Extracts-June-2018.pdf

Urban innovation for liveable cities. A holistic approach to sustainable city solutions. (2016). Think Denmark. White papers for a green transition. State of Green 2016. Retrieved from www.stateofgreen.com

Velenturf, A. P. M., & Jensen, P. D. (2015). Promoting Industrial Symbiosis: Using the Concept of Proximity to Explore Social Network Development. Journal of Industrial Ecology. Retrieved from http://epubs.surrey.ac.uk/812461/1/Complete%20thesis%20Final%20with%20viva%20corrections_submitted%20September%202016.pdf

Zaman, A. U. (2014). Measuring waste management performance using the "Zero Waste Index: The case of Adelaide, Australia. *Journal of Cleaner Production*, *66*, 407–419. Retrieved from https://espace.curtin.edu.au/handle/20.500.11937/54995

Chapter 4
Circularity and Cultural Heritage Stock

ABSTRACT

With the development of a modern designer's workshop, various smart city issues have to be included in line with more conventional analyses. Presently, we also face emerging circular economy theme, which has a high impact not just on the introduction of circular loops into the flow of building materials, but also on the design approach and management choices. Historic heritage buildings should also be considered within this new theme. Most of the existing research either deals with new or modernized buildings, or with the re-use flows of various materials, often coming from historic buildings gone outside the limits of repair. This chapter explains the proposed approach and includes case studies where such an approach has been provided.

CIRCULAR APPROACH IN CASE OF RETROFITS

Building sector is one of the key consumers of energy in every continent and Europe is no exception. Throughout a number of years, EU has enacted several directives dealing with energy efficiency in buildings with the aim to reduce their energy use. Unfortunately, these mainly concern modern stock, whereas representatives of the Architectural Heritage have to be specially treated during each design process. Countries adapt their own rules, which often differ from region to region. Furthermore, the European Union Treaty does not include the Cultural Heritage within legislation, which would allow

DOI: 10.4018/978-1-7998-1886-1.ch004

bridging the gap between historic buildings and energy as well as sustainable issues within retrofit processes. The former by now has become an element deeply integrated in everyday design. With the development of a modern designer's workshop, we now have to include resiliency, passive, ecologic, plus energy, nZEB and other issues (Dalla Mora et al. 2015). Currently, when the civilisation is confronted with a barrier of source scarcity, a shrinking level of not just fossil energy sources but other sources as well, there is a need to remodel the curricula of design education and move first from linear autonomous solutions into interdisciplinary and then circular ones (Rynska 2016). This different approach to design process should include use of external environmental parameters specific to each given location. It is also strongly connected with the emerging circular economy theme. Nevertheless, a deeper analysis distinguishes certain characteristics (Francesca 2017), (Rynska et al 2018). First, development of societies and urbanization should be consistent on a level deeper than presently, and be included within the design processes, organization and planning, as well as modernization and redevelopment procedures of existing urban tissue. Second, urbanization process should be perceived holistically, as an interaction and harmonious development of both natural and manmade environments, with solutions based on the best technical and technological standards available, and circular economy choices. Last - described ideas are achievable only, if we include continuous cooperation between urban planners, architects, specialist consultants, as well as energy effective interdisciplinary solutions to achieve efficient energy measures. One of the thresholds is circular economic feasibility; the other is health and wellbeing of the users whose needs should always be discussed as a priority to any other solutions. Social and education issues may also form important barriers.

Sustainable conditions for design and construction of buildings directly involve implementation of particular design procedures and management methods, specifications of environmentally friendly building materials. Designers participating in building design represent a group of professionals who have direct influence on the process where the building materials become part of a circular closed loop economy. Such approach reduces the dependence of further economic development on the finite supply of natural resources. The choice and the management system is a complicated procedure in which each phase of the building's life is analysed - it includes production process, used building techniques and technologies, expected building's end of life which should include re-use of initially build-in either materials or structural elements (Kozminska et al. 2018).

The EC introduced Circular Economy Package in 2015. It contained basic definition of the circular economy as the approach where economic value and the path to further economic development a low-carbon society are united in a single policy approach. Adoption of necessary policies also means a more sustainable footing and transformation towards more efficient and circular business (EC 2018), (Pomponi et al. 2016). This in turn requires radical change compared to green-filed projects, which are the most common investing choice (Sanchez et al. 2018). The new principles include product recovery, management, life cycle assessment (LCA), design for disassembly sequence planning, adaptability, closed material loops and deconstruction. These very general indicators do not include any particular technologies when choosing between green field construction and adaptive reuse (Sanchez et al. 2018), and between optimization of adaptive reuse or complete reuse of existing structures. It is even more difficult when considering Cultural Heritage buildings that do not have economic value – but rather draw the attention to the cultural, aesthetic, social and historical issues (Rypkema 2008).

The modern circular economy is based on overwhelming saturated markets not scarcity (Baker-Brown 2017). The industrial revolution initiated in 18th Century, was based on linearity and a certain type of efficiency allowing production of more and of better quality goods. Some researches implicate that there is a direct connection between metabolism and circulation (Heynen & Kaika & Swyngedouw 2006), but is should be noted that this assumption is based mainly on the circulation of capital as value in motion perceived from the social angle. Nevertheless, the idea that urbanization is not only a technical issue, but to a large extent is based on the policies similar to social processes, is a correct one (Cook & Swyngedouw 2012). Prevailing contemporary social expectations, which support unlimited consumption in major part, are responsible for the most visible civilization outcomes including large volumes of non-used waste often collected as landfill and overconsumption of natural resources. Regardless whether the idea of circular economy is perceived from a social or technical angle, it should be noted that the transformation from a linear to a circular development is not an easy process and only a limited progress has been accomplished so far. This is due largely to the low consumer interest and awareness (Kirchherr et al. 2018).

Issues that are included in the transition zone leading towards circular economy may be allocated into phases initiated by the useful application of materials, extended lifespan of products and parts, to be concluded with a smarter product use and manufacture (White Paper 2018). The idea of re-using of goods, components and materials with the main aim – to preserve the value of existing building stock and their aesthetics can be traced in the sub-areas such as (White Paper 2018):

- **Recycling**: Where the materials are processed to obtain the same or lower quality products.
- **Repurposing**: Where discarded products or parts may be used in a new product with a different function.
- **Refurbishment:** Where old products are restored and brought up to date.
- **Repair:** Concerned with repair and maintenance of defective products, to be used with its original function.
- **Reuse:** Where another consumer may reuse discarded product, which is still in good condition and fulfils its original function.

These sub-areas are quite visible when applying within the modernization and preservation procedures of a building stock. Already there are many examples of deep retrofit procedures, which outside preservation of existing building materials also achieve efficient energy solutions. Norwegian building Kjørbe (Figure 1 and 2) may be pointed out as one of good examples, where high emphasis was used on the choice of environmentally responsible building materials including re-use of former external façade elements as internal office partitions (Klimowicz 2018). It is one of examples confirming that in construction business, circular economy manages and reuses manufactured capital, multiplying the service of goods and lowering the waste's volume (Baker-Browne 2017). Hence, contemporary building industry faces a triple challenge: development of efficient and waste free construction methods, designing buildings for minimum resource consumption during operation and maintenance, as well as developing methods allowing for deconstruction while preserving the highest value (Baker-Browne 2017).

It should be noted that in most papers published so far, the issue of circular economics used in retrofit historic buildings is rare. Even the IDSA Principles of Design for (IDSA) appear to miss this area, except maybe for a point concerning use of recycled materials.

Since historic buildings constitute at least 25% of the European built environment they have a major role to play in delivering CO_2 emissions reduction targets along with the rest of the domestic stock out of which more than 50% is more than 50 years old (Moran et al. 2012, 2014), (Ginks et al.

Figure 1a. Interior of Norwegian Kjørbe Building in Oslo, office units with walls from renovated former façade elements

Figure 1b. Interior of Norwegian Kjørbe Building in Oslo, free space office area with semi-private zones screened with vertical made from pre-used plastic bottles
Source: (photo: Rynska E.)

Figure 2. Interior of Norwegian Kjørbe Building in Oslo with semi-private zones and ceiling screened with vertical made from pre-used plastic bottles
Source: (photograph: Rynska E.)

2017). The current EU interventions often include criteria inapplicable to historic buildings. This state is fostered due to already mentioned status quo where the regulations concerning preservation of architectural heritage and rational use of energy are not linked, and also due to the fact that energy retrofitting standard to be used in historic buildings is still missing (Litti et al. 2013). In some countries, research organizations have been developing strategies at evaluating energy efficiency solutions in historic buildings and offering a set of guidelines. Such examples may be found in Italy (Di Ruocco et al. 2016), where the guidelines show a range of architectural restoration projects involving the use of innovative, less invasive more efficient building materials adaptable to the historic contents (MiBACT 2015).

Historic buildings have significant cultural value and were built with technologies and materials that promote fabric breathability and may be perceived as part of the circular economy approach. Especially, if used strategies will allow to maintain their location or, re-use historic building materials and structural parts in other cultural heritage buildings when upgrading their substance (Rynska 2008). Some decisions have an impact on both existing fabric and aesthetics, therefore prior to any design process which deals with such cases, there should be a requirement to perform a cultural diagnosis of each particular site in order to establish the sequence of procedures. It should be noticed that this approach is an emerging attitude, which is being used simultaneously with the growth of increasing attention to preserve architectural heritage where typical invasive retrofit interventions are not applicable. The need for a replicable methodology to improve the sustainability of historic buildings based on the integration of energy efficiency solutions with renewable technologies (Pisello et al. 2015), (Webb et al. 2017), (Rodrigues et al. 2017) and the introduction of circular economic loops is one of the major subjects which should be considered. In view of this background brief, it should be noted that proceeding urbanization and the impacts of climatic changes are pressing the cities to prepare new development procedures. One of those paths is circular economy (Prendeville et al. 2018) with the idea of the re-flow of resources through economies for consecutive re-use. Whereas this concept is currently on an experiment level, in many ways it can be applied to the cultural heritage stock under the condition that circular issues are merged with the cultural diagnosis. Since there is a there is a lack of consensus on what a circular city should constitute, this proposition might be one of the possible choices.

Every approach to the retrofit design process is faced with a decision making process attributed to the best case selection depending on the initially set goals (Kass et al. 2016). Therefore, development of evaluation criteria, which will form a set of red flag choices, filed down to the specific building's case should be foreseen. In case of historic buildings, the procedure is often very complex and therefore there is a requirement to form a net of procedures where recycling, repurpose, refurbishment, repair and reuse will form a basic choice pattern from which further choices and possible management procedures can be undertaken. This net of procedures should be also supported by the city policy and social requirements. Other researchers have also followed this line of thought, where scientific approach also includes a variety of framework conditions (Kass et al. 2016).

This Chapter presents possible approaches to the cultural heritage forming part of city urban tissue when the main aim is not only historic diagnosis of the most valuable building elements, but also re-use of existing valuable materials either in-situ or as a part of other revitalized buildings dating back to the same period. This is in line with the idea of circular development for the city areas introduced by the EU in 2015.

Case Study: Housing in Mokotów District, Warsaw, Poland

In 1897, a German investor Georg von Narbut, bought a large parcel of land located just outside Warsaw's limits and started to divide it into smaller plots and streets. A large market square was foreseen half way down the main street. The investigated building, although not listed for its special architectural interests (Figure 3), is located in an area currently under historic urban preservation. This zone is subject to restrictions as to street layout and some of the buildings are listed in the Register of Historic Monuments. Most of the volumes in this area were built in mid-half 20th Century. Since a number of documents were lost during the Second World War, the official existing data placed the construction date of the studied case in early -50-ties of the previous century, in a style described as early Modernism, with a simple plain façade devoid of ornament. This was the information given to the new owner of the estate, who approached a designing team to prepare a proposal. According to the Client's brief, proposed interventions were to be divided into two categories:

- Recovery and rehabilitation, which included extension of the building by two floors; providing adequate insulation and waterproofing, introduction of new technical systems; achievement of energy class B certificate, removal of all internal finishing details and partition walls.
- Aesthetics, which included redevelopment of the façade including provision of larger window openings (as the level of daylight in some of the rooms did not meet present Polish Building requirements); modernisation of the balcony balustrades and introduction of new features which included new external cladding elements and a new colour scheme.

This initial brief was consulted with the representatives of the Warsaw Historic Heritage Preservation Department, who contested some of the indications and requested preparation of a cultural and historic diagnosis of the site with the main aim to disclose the original colour of the building's façade, which in general was to be off-white. Research was conducted in two areas. The first one consisted of archive research. This allowed defining primary architectonic solutions and later-on choosing the location of the site openings in characteristic areas. The samples were taken on-site as cross cuts containing stratigraphic layers of the previous internal and external finishing. Beside estimating initial colours and finishing materials of the facades,

Figure 3. View of the existing state (2018) of the Mokotów building off the Square
Source: (photo: Rynska E.)

research included façade's primary design features used in the main entrance area, which included loggias on the first and second floor level, as well as confirmation of the historic portal's initial form. Other secondary themes included analysis and samples from the building, chosen in characteristic points which might have had impact on the building's cultural value. These included – main entrance hall, staircase, ceramic tiles dating back to beginning of 20th Century and window frames. The outcomes proved that the history and the value of the building were different from the initial assumptions, as it was in fact constructed in 1934. There is no name of the designer mentioned in any of found documents, but historic details belong more to the earlier period (1925-27), when the oldest building in the vicinity was built. Possibly, this unusually long pre-development period caused the mix of two different architectonic styles – Historic and Modern. Exposed corner site was flanked with a large balcony off western façade, and an open loggia facing the square. The assumption was that one of the floors was to remain in the owner's hands, whereas remaining apartments were destined to be let. There was a small retail in the semi-basement area (Rynska et al. 2019).

The building was raised, from earlier purchased building elements (it was evidenced that the existing window frames were marked with 1932 date). In 1945, the building was burned out during the war siege, but the external façade, staircase and some of the internal fit-out remained. It was re-built in 1945-47 by the owners, but some of the works were still being conducted in the 1950. In 1955, as many other estates, it was taken over by the Treasury of State, which was further confirmed by the entry into the Land Register in 1956. Building became part of the communal property stock, which meant that it was very badly maintained. Even after many years of misuse some of the original features remained – such as conch shaped niches for tiled stoves in the dining rooms, elliptical entrance hall leading into staircase area still clad in original terrazzo, and original early 20th Century floor ceramic tiles in some of the rooms (Figure 4). Prepared historic analysis lead to a conclusion that the main historic and cultural value is the volume of the tenement building – with a height lower than other, later build typical modernistic buildings, but similar to the no 50 plot where the historic Neo-renaissance building designed by arch. T. Zielinski stayed intact.

Early 60-ties brought new investments throughout Warsaw, and the adjacent plots were invested with a building that was located right against the balcony wall – and a building off the gable wall in the north side completing the square's façade. This solution formed a new situation where the building had only two external walls, whereas remaining (dwindled ones) became walls facing internal courtyard which was entered through a gate opening accessed off a major street. Analysed building remained a tenement building throughout the time. Some of the apartments were divided into smaller flats.

The first Energy Directive brought changes throughout Polish cities. Many of the buildings were given permit to provide external insulation in order to achieve better energy conditions including the one under consideration, which was given such permit in 2000. The works were conducted in 2005. As presently, all facades have external Styrofoam panels, it was impossible to conduct a complete diagnosis. Nevertheless, it can be stated that the form of the façade was subjected to the rhythm and logic characteristic to the Historicism tradition (Neo-classical style) – presenting historic modernism stylistics. There is no trace of photographic evidence documentations disclosing how the building looked prior to modernization – even though such evidence should form a part of the deep insulation design. Warsaw state archives included an accepted version of the new facades, but differences were found between the accepted forms and colour of facades and the ones that were actually built. The contents describing provision of insulation works indicate that all architectonic articulations (even those more than 3.0 cm deep) had to be covered in layers of Styrofoam in order to receive a flat façade surface. In the areas where historic steel balcony balustrades were imbedded in the external walls, the contractors were requested to cut off side flower baskets, destroying the original design. Stratigraphic analysis of the plaster layers'

Figure 4a. A view of the original early 20th Century balustrade with original terrazzo slab on the staircase landing

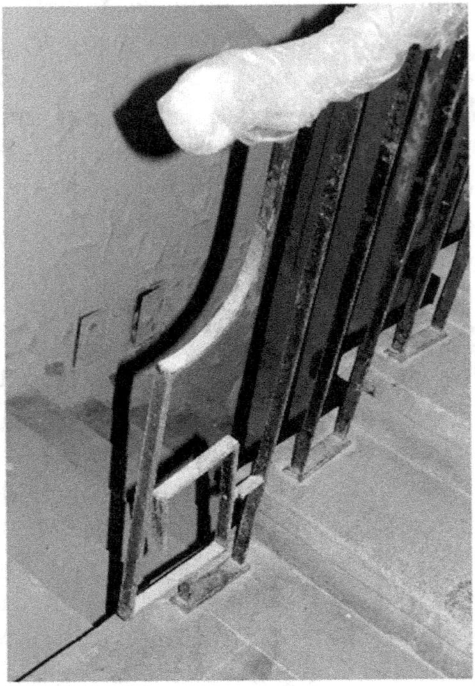

Figure 4b. Original ceramic tiles
Source: (photo: Rynska)

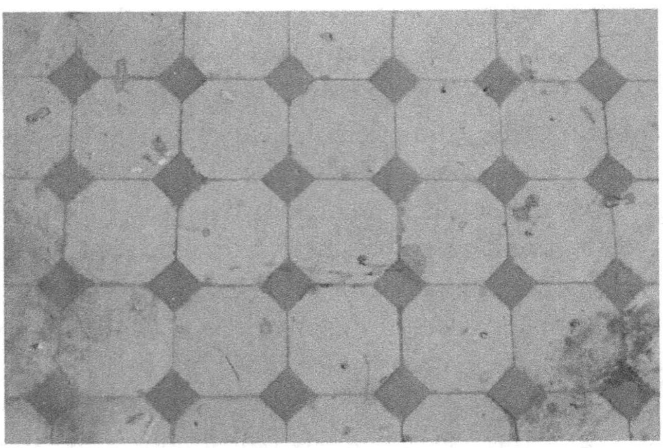

Figure 5a. View of the reinstated façade – 1&4 first and send floor loggias surrounded by deep portal frame this area possibly in part finished in grey plaster imitating stone, 3 – main façade in light salmon pink stucco, possibly with crème vertical articulations, 5 & 2 – steel balcony elements according to original locations, 9 – original historic entrance portal
Source: (Rynska et al. 2019)

samples and a thorough search on the colours used at the time for Warsaw's tenement buildings disclosed that the initial main colour of the façade was light salmon pink stucco.

According to the technical description of the façade, when covering with the insulation panels, the standard of the existing plaster was very bad, black from the fire and with uneven often-loosened surface. Hence, after deep insulation process original plaster was destroyed beyond recovery. Even if the insulation cladding were to be removed, a reconstruction of a new plaster layer would have to include materials and techniques not accessible on a contemporary building market.

The main problem with the insulation works is that they have destroyed the main architectonic, urban and cultural values of a building located in a very exposed area. Site openings, as well as research of other buildings built

Figure 5b. Possible colour scheme of reinstated façade, research proves that the original door has vertical window element in order to make the oval hall entrance brighter
Source: (Rynska et al. 2019)

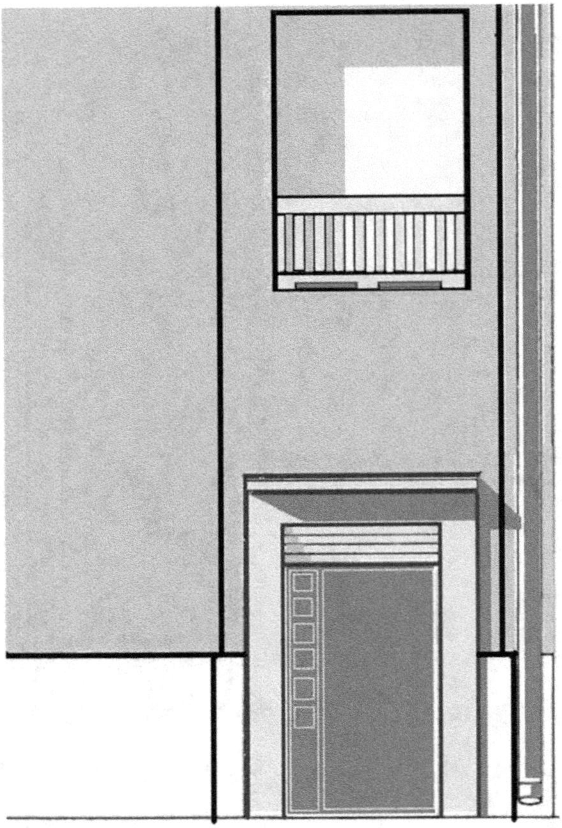

in Warsaw at the corresponding period, allowed the researchers to prepare, a proposition of the original frontage facing the square (Figure 5a)

In case of discussed building author's proposals aims to arrive at a balanced solution, which will include existing state of the building and re-introduction of its historic values. Hence, it was proposed to revitalize pre-war most important features. These being: the door portal and composition of surrounding detail including loggias, maintaining original steel balcony balustrades and the original articulation of facades. It is proposed to reinstate the three dimensional effect of the façade through application of such details as a socle, pilasters, the crowning cornice and vertical pilaster strips surrounding the portal's area. The present Master Plan conditions require the extension of the building by two levels to the height of surrounding buildings. Therefore, the conclusions indicated that the crowning cornice of the building should act as a visible element – a joining detail between the old building and the new extension. The open loggia including the balustrades on the Eastern façade should be possibly reinstated, as this feature forms a uniform solution with the main entrance portal – the façade's main ornament – which should be uncovered from the Styrofoam layer and the original grey stucco reintroduced. Research does not prejudge whether the former balustrade can be uncovered from the insulation and then undergo restoration, or if a newer version of the balustrade has to be constructed. It may be possible to maintain the newer Warsaw tradition and close the opening in a form of a porte-fĕnetre window mounted behind the balustrade. Existing floor ceramic tiles dating to prior 1939 (1st and 3rd Floor) should be either kept in situ or located to a more exposed area. Main staircase area, including all original elements – balustrades, terrazzo, and form of the steps should be renovated. This solution should be applied also to the reinstatement of the main entrance off the square area.

Cultural diagnosis had a major influence on the originally proposed design. Whereas the initial main line of thought was to keep as little as possible of the original substance and exchange it with modern high quality building materials, later the Client requested to maintain as many of historic features as possible. Such choice is in fact often very rare, unless the building is well known as a historic icon and under strict historic monument regulations. In case of historic buildings that are located in the historic urban areas, but not under the historic regulations – developers mostly continue more in line with the financial gains, not the true value of the solutions which when revitalized might work to the benefit of the local communities and fulfil circular economy expectations. This new approach allowed to introduce certain elements dealing with circular economy side of the building's management. The form of the building will remain unchanged and most of the structural and building elements either re-used or repurposed. Reinstatement of the loggias and use of porte-fênetre is foreseen, as well as the façade's original colouring. Building's extension will be moved back from the line of the original façade. The height of the new building will correspond with the surrounding, as this choice is has to comply with the Master Plan conditions, and is in line with traditional economic approach. Existing building materials and architectonic features, including steel balcony balustrades, will be kept in place. The topmost level will be constructed in lightweight mostly glass façade elements, as to achieve different aesthetics from the older part of the building. The staircase remains unchanged both where dimensions and external finish is concerned. These choices prove that it is possible to introduce circular economy features when modernizing historic substance allowing for a synergy achievement of re-use and repair of existing elements under the historic and cultural guidance.

Case Study: Higher School of Ecology and Management (Wseiz), Warsaw, Poland

This particular case study is actually located in Warsaw, in the same area as the former Narbutta case. Ornamental and Forged Steel Plant Company "Hammer" located at 12, Olszewska Street was constructed in the 1920. At its prime, it manufactured balustrades, fences and stairs. After the war, function was changed to a gun-smith workshop and then a printing house. Forgotten, for many years it became a home for local numerous flocks of pigeons. Since 2010, based on Archimed Architects design of 2005, it houses students of the Higher School of Ecology and Management. Building presents many typical features of a early 20th Century small manufacturing plant, with red brick façade and large steel framed windows. Designers decided to reinstate the original pre-war outlook of the building. Unfortunately, former factory was in a sad state of technical condition, walls ware soaked with rainwater and many of the structural elements did not fulfil present load bearing expectations. During construction process, in many cases structural elements had to be re-designed on-site, as site excavations proved that the existing technical survey did not meet neither reality, nor did it mirror the technical state of members. Architects had to place external insulation with ceramic cladding (Figure 5b). The choice of the investor was to keep most of the existing building structure and as many characteristic features as possible. Architects were therefore faced with a difficult choice, especially when designing façade features. Since existing façade bricks were of low technical value and the re-use of materials from other sites was, and still is, a very complicated process in Poland, after consulting the owner, it was decided to use an elastic ceramic finishing layer. This allowed to maintain the most important feature – the 19th Century industrial brick outlook of the front facade which gave the atmosphere of reminiscent past – even if modern building materials were actually used.

Small dilapidated side buildings build after 1945 were demolished, a new educational wing designed in its place. This new extension does not mimic the older building's architecture features and is very distinct with its white façade from fiberglass-cement panels and a very modern outlook. Historic building was extended in steel and glass structure, by one floor holding both offices and classrooms. This extension was also planned to have an industrial lightweight feeling which does not compete with the main building's features. Existing historic forged steel elements were renovated and replaced on the façade in their former locations. Interiors are sparse with industrial character maintained by untreated brick, use of architectonic reinforced concrete walls and visible installation ducts (Figure 6a and 6b). Even though there was no request to provide a cultural diagnosis, the owner requested to have such a document which was an additional support document during design process. It should be concluded that this merge of historic and modern architecture suits the school's education technical and artistic profiles.

Case Study: Koszyki, Warsaw Downtown Area, Poland

Koszyki (The Baskets) complex was a third shopping and trading area constructed in Warsaw at the break of the 20th Century. This large market hall introduced better organisation of trade procedures, and much better hygienic conditions. Building was constructed in a location initially used as an open market square established in this area for the inhabitants of the south part of Warsaw. It was built by the Warsaw government on a land purchased from a private owner. This complex was constructed in 1906-1909 in a secession style according with a design prepared by architect J. Dzierzanowski, who at the time was employed by the City Management Construction Department. The lay-out is formed on a letter "C" opening towards the main street – Koszykowa (circa 77 x 15m). Building was constructed by Kuksz & Lüdtke construction firm, and the steel riveted load-bearing structure was prepared by W. Gostynski enterprise. Steel hardware, including gates were manufactured by Steel Manufacturing Plant owned by H. Zielezinski. Sculpting ornaments were authored by Z. Otto, and include one of few of Warsaw examples where animals were included as part of sculpture. Other details included flower bouquets and fruits. This building was considered at the time as being very modern. Pedestrian entrances were located in the side wings, whereas tradesmen were accepted through a gate located in the main part of the building. Market hall included four large two story shops in the front part, twenty four smaller shops, sixty meat stands, twelve fish stands and a hundred and forty four other stands. Each stand was designed

Figure 6a. Main entrance to the Higher School of Ecology and Management, Warsaw

Figure 6b. Main entrance hall to the school designed in modern technology maintains the post-industrial atmosphere of the building
Source: (photograph: Rynska E.)

individually i.e. those where fish were sold were fitted out with small tanks, and those where meat was sold had marble slabs. Each of the vendors had access to underground cool storage. Trading took place both before as well as behind the market building, where also an administration building with a flat for the manager was built. The main fault was insufficient area to park vehicles and difficult access from narrow Koszykowa Street. Market hall was burned down and in a large part destroyed during the Warsaw Uprising in 1944. After the Second World War remains were taken over by public-private owner Cooperative „Spolem". After 1950, building was extended with low pavilions surrounding the internal courtyard and used as external market places. In 1965 building was registered in the Register of Historic Monuments which did not stop the owner from planning a total modernization. It was to be demolished and a new modern glass building constructed on the site (Zielinski 2000). Adequate permissions were actually issued. In 2006, building was sold to a private developer firm Quinlan Private Golub, who requested an expert opinion according to which building was in a state of near-collapse which led to the removal of trade and immediate shut down. This action was met with many protests both from the architects and historians, as well as from Warsaw inhabitants voting against demolishment of historic buildings. The main argument was the requirement to maintain the atmosphere of this historic area. These actions were supported by a decision issued by the Warsaw Officer of Historic Monuments containing suspension of all works. It was only then that the investor decided to maintain the historic building or rather re-build it, with the use of existing steel structural elements and reinstatement of architectonic elements. Koszyki changed hands once more, and in 2009 Avestus Real Estate bought the building and demolished it leaving only the two side entrances and the main walls. Investor decided that the main load bearing steel elements were to be renovated and placed once more in the modernized building. According to the plan accepted by the Main Historic Officer, structural elements were to be placed in location after finalisation of the new underground parking level. Since these expectations were no met by the investor, construction was again halted. In 2012, half-finished building was purchased by Griffin Groups who accepted revitalisation procedures proposed by the city officials and hired JEMS Architects to provide a new set of design drawings. New building was opened in October 2016. Most of the original steel structure load bearing elements were re-instated, secession ornaments and original shop signs restored (Figure 7).

Original ceramic tiles were re-laid in the entrance halls, steel structure elements were painted to its original shade of green (Figure 8). According

Figure 7a. Re-built side entrance wing to Koszyki Market Hall, with original facade ornaments, new office building visible on the right

Figure 7b. Enlargement of the entrance portal with renovated sculptures and ornaments
Source: (photograph: Ryńska E.)

Figure 8. Renovated steel riveted structure placed in its location and painted to the original green colour, original floor tiles are visible in the back
Source: (photograph: Ryńska E.).

with the investor's concept building was to include both the trade and food court functions. Level -1 was taken over by a market store, and -2 destined as an underground parking lot. Two modern 6 level office buildings were built as side wings on both sides of the historic building.

Maintaining the historic location and urban layout of the city is one of the most important features for the cultural identity of the city. A place is only a fragment of cultural space, which is subconsciously given certain meaning during the course of its creation (Rynska 2008: 334-335). If the meaning of the place is lost, then the history and identity is also lost, unless it is returned based on the cultural and historic analysis of the site. In order to understand the meaning of a site, we have to understand a code of the language presented in a material form or space. This code is also part of the cultural background for the people belonging to a certain part of civilization and is a part of heritage. This principle can be also described as the spirit of a place – something that in the double case of Mokotów buildings - has been entirely lost. In order to revitalize the identity of the place designers used various different features, both where the external façade as well as public areas were concerned. One of the weakest points in contemporary urban fabric is disappearance of a composition factor in the modern urban planning and absence of acceptance that new solutions – circular economic, sustainable and smart city issues should form part of the new design. Therefore, in order to achieve harmonious urban composition, owners and designers should meet more than just legal requirements and even the newest construction solutions.

In contemporary world, the attitude of investors, when embarking on a development, which includes historic buildings, is to find the least troublesome solution from their point of view. In most cases, this means dismantling of existing building's substance and creating a new one in its place. If pressed towards circular economy choices, the easiest choice will be recycling with the aim to re-use products on different sites. Therefore, there appears a requirement for additional reference documents specifying what are the best circular choices to be used as there is no evident indication when and what buildings should be dismantled or treated with different procedures. In case of buildings located in cultural heritage zones, cultural diagnosis might prove to be a major issue. The second, more developed choices are repurposing and refurbishment, where discarded elements may be used in new buildings with a different function, or in case of a historic building - relocated within the building in order to enhance the most valuable features. Again, circular economy procedures do not refer to any additional documentation on which the decisions could be made. As to which parts can be repaired, removed and relocated while still maintaining heritage and circular values – there is also no indication as to how to proceed. The final possibility is the use of discarded historic building elements on a different site. The assumption is, that in such cases a cultural diagnosis should indicate where those elements should be reused in order to enhance their value.

CONCLUSION

Cultural heritage is part of a very important area. It shows who we were and is a key element of our cultural past. Yet, 20th Century approach has marked such buildings more as a hindrance than an asset. Unfortunately the same phenomena appears where circular economy is concerned. It should only be hoped that historic buildings will not end up as part of the re-use and recycle processes. Our society is economy driven and when faced with the requirement to move towards circular economy choices without understanding the basic requirements, within the construction the easiest choice will be recycling which is only one of the possible routes. Therefore in case of historic buildings and historic complexes there is a need to provide additional document, which in this chapter has been described as cultural evaluation.

REFERENCES

Baker-Brown, D. (2017). *The Re-use Atlas. A designer's guide towards a circular economy*. RIBA Publishing.

Cook, I. R., & Swyngedouw, E. (2012). Cities, Social Cohesion and the Environment: Towards a Future Research Agenda. *Urban Studies*, *49*(9), 1959–1979. doi:10.1177/0042098012444887

Dalla Mora, T., Cappelletti, F., Peron, F., Romagnoni, P., & Bauman, F. (2015). Retrofit of an historical building towards nZEB. *Energy Procedia 78*. Retrieved from https://www.researchgate.net/publication/290022096_Retrofit_of_an_Historical_Building_toward_NZEB

Di Ruocco, G., Sicignano, C., & Sessa, A. (2016). Integrated methodologies energy efficiency of historic buildings. *Procedia Engineering*. doi: 10.1016/j.proeng.2017.04.328

EC. (2011a). *Territorial Agenda of the European Union 2020 — Towards an Inclusive, Smart and Sustainable Europe of Diverse Regions*. Agreed at the Informal Ministerial Meeting of Ministers responsible for Spatial Planning and Territorial Development on 19th May 2011 Gödöllő, Hungary. Retrieved from http://www.eu-territorial-agenda.eu/Reference%20 Documents/Final%20 TA2020.pdf

Francesca, R., Oberegger, U.F., Lucchi, E., & Troi, A. (2017). Energy Retrofit and Conservation of a Historic Building Using Multi-Objective Optimization and an Analytic Hierarchy Process. *Energy and Buildings, 138*, 1–10. Doi:10.1016/j.enbuild.2016.12.028

Ginks, N., & Painter, B. (2017). Energy Retrofit Interventions in Historic Buildings: Exploring Guidance and Attitudes of Conservation Professionals to Slim Double Glazing in the UK. *Energy and Buildings, 149*, 391–99. doi:10.1016/j.enbuild.2017.05.039

Heynen, N., Kaika, M., & Swyngedouw, E. (2006). *In the Nature of Cities. Urban Political Ecology and the Politics of Urban Metabolism*. Routledge, Francis Taylor Group.

IDSA. (n.d.). *Ten Principles of Design for Environment – Industrial Designers Society for America (IDSA)*. Retrieved from http://www.idsa.org

Kass, R., Caffo, B. S., Davidian, M., Meng, X.-L., Yu, B., & Reid, N. (2016). Ten simple rules for effective statistical practice. *PLoS Computational Biology, 12*(6), e1004961. doi:10.1371/journal.pcbi.1004961 PMID:27281180

Kircherr, J., Reiker, D., & Hekkert, M. (2017). Conceptualizing the circular economy: An analysis of 114 definitions. *Resources, Conservation & Recycling*. Retrieved from https://www.researchgate.net/deref/http%3A%2F%2Fdx.doi.org%2F10.1016%2Fj.resconrec.2017.09.005

Klimowicz, J. (2018). Chosen case studies of nZEB Retrofit Buildings. In *Design Solutions for nZEB Retrofit Buildings*. IGI Global. doi:10.4018/978-1-5225-4105-9.ch009

Kozminska, U., & Rynska, E. (2018). Materiały konstrukcyjne i budowlane spełniające zasady zrównoważonego rozwoju. In S. Firlag (Ed.), Zrównoważone budynki biurowe, PWN Editing House in cooperation with PLGBC. Warsaw: Academic Press.

Litti, G., Audenaert, A., & Braet, J. (2013). Energy Retrofitting in Architectural Heritage, Possible Risks Due to the Missing of a Specific Legislative and Methodological Protocol. In *The European Conference on Sustainability, Energy and the Environment 2013 Official Conference Proceedings*. IAFOR The International Academic Forum. Retrieved from https://www.researchgate.net/publication/265864668_Energy_Retrofitting_in_Architectural_Heritage_Possible_Risks_Due_to_the_Missing_of_a_Specific_Legislative_and_Methodological_Protocol

MiBACT. (2015). *Guidelines for energy efficiency improvements in the cultural heritage – Architettura, centri e nuclei storici ed urbani*. MiBACT. Retrieved from http://www.beniculturali.it/mibac

Moran, F., Blight, T., Natarajan, S., & Shea, A. (2014). The Use of Passive House Planning Package to Reduce Energy Use and CO 2 Emissions in Historic Dwellings. *Energy and Buildings, 75*, 216–27. Retrieved from https://www.researchgate.net/publication/260993229_The_use_of_Passive_House_Planning_Package_to_reduce_energy_use_and_CO2_emissions_in_historic_dwellings

Moran, F., Sukumar, N., & Nikolopoulou, M. (2012). Developing a Database of Energy Use for Historic Dwellings in Bath, UK. *Energy and Buildings, 55*, 218–26. Retrieved from http://isiarticles.com/bundles/Article/pre/pdf/67302.pdf

Paper, W. (2018). *Circular Economy in Cities. Evolving the model for a sustainable urban future*. World Economic Forum REF 260218 — 00034436. Retrieved from http://www3.weforum.org/docs/White_paper_Circular_Economy_in_Cities_report_2018.pdf

Pisello, A. L., Petrozzi, A., Castaldo, V. L., & Cotana, F. (2014). On an Innovative Integrated Technique for Energy Refurbishment of Historical Buildings: Thermal-Energy, Economic and Environmental Analysis of a Case Study. *Applied Energy, 162*, 1313–22. Retrieved from https://www.sciencedirect.com/science/article/pii/S0378778818313537

Pomponi, F., & Moncaster, A. (2016). Circular Economy for The Built Environment: A Research Framework. *Journal of Cleaner Production, 143*, 710–18. Retrieved from https://www.repository.cam.ac.uk/handle/1810/261963

Prendeville, S., Cherim, E., & Bocken, N. (2018). Circular Cities: Mapping Six Cities in Transition. *Environmental Innovation and Societal Transitions, 26*, 171–94. Retrieved from https://www.sciencedirect.com/science/article/pii/S2210422416300788?via%3Dihub

Rodrigues, C., & Freire, F. (2017). Adaptive Reuse of Buildings: Eco-Efficiency Assessment of Retrofit Strategies for Alternative Uses of an Historic Building. *Journal of Cleaner Production, 157*, 94–105. Retrieved from https://www.researchgate.net/publication/316354869_Adaptive_reuse_of_buildings_Eco-efficiency_assessment_of_retrofit_strategies_for_alternative_uses_of_an_historic_building

Rynska, E. (2008). Rehabilitation and adaptive reuse of historic buildings in Poland. *WIT Transactions on Ecology and the Environment, 113*, 327-335.

Rynska, E. (2016). Interdisciplinary training within education curricula for architects and engineers. *Global Journal of Engineering Education*, 202-206.

Rynska, E., Kozminska, U., & Rucinska, J. (2018). Effectivity–ecosphere–economics in nZEB retrofit procedures. *Environmental Science and Pollution Research International*. doi:10.100711356-018-2446-8 PMID:29936610

Rynska, E., & Lewicka, M. (2019). Closed circulation loops in historic buildings. Cultural diagnosis as one of the major factors in a contemporary designer's workshop. *Urban Development Issues, 61*, 41-50. Retrieved from https://content.sciendo.com/downloadpdf/journals/udi/61/1/article-p41.xml

Rypkema, D. D. (2008). Heritage Conservation and the Local Economy. *Global Urban Development, 4*(1), 1–8. Retrieved from http://www.globalurban.org/GUDMag08Vol4Iss1/Rypkema PDF.pdf

Sanchez, B., & Haas, C. (2018). Capital Project Planning for a Circular Economy. *Construction Management and Economics, 36*(6), 303–12. Retrieved from https://www.tandfonline.com/doi/abs/10.1080/01446193.2018.1435895

Webb, A. L. (2017). Energy Retrofits in Historic and Traditional Buildings: A Review of Problems and Methods. *Renewable and Sustainable Energy Reviews*. Retrieved from https://www.researchgate.net/publication/316653158_Energy_retrofits_in_historic_and_traditional_buildings_A_review_of_problems_and_methods

Zieliński, J. (2000). Atlas dawnej architektury ulic i placów Warszawy. Tom 6. Kępna–Koźmińska. Warszawa: Biblioteka Towarzystwa Opieki nad Zabytkami.

KEY TERMS AND DEFINITIONS

Cultural Diagnosis: A document sometimes provided in Poland for historic buildings and urban sites, it includes both historic and technical transformation data. If required content can be deepened with technical analysis of structure and on-site openings and samples extracted in order to i.e. prepare stratigraphic analysis and check original cladding and colour.

Chapter 5
How to Prepare a Circular Brief for a Designer?

ABSTRACT

Since this is a beginning of a new holistic approach, one of the most difficult issues is preparation of a circular designer's brief, which will be helpful at all phases of design, construction, maintenance, end of building's life, and various types of material loops. Currently there is no single approach. Basically, it exists more as a general attitude statement and still has to be developed. This chapter will include some thoughts and possibilities as to the design approach and management that can be used for various different projects. Some design outcomes from studios run by the author jointly with members of Faculty of Architecture staff are included. These deal with some circular solutions proposed to the students for further development. Two case studies presented in the appendices will prove that circular approach should be used in various harsh environments, as well as city conditions.

TEMPORARY DESIGNER'S CIRCULAR WORKSHOP

As already mentioned the approach towards our contemporary civilisation development is based on abundance not on hindered access to resources. Emerging circular economy appears to be more sustainable than the existing linear economy as it is based on regional resources and uses less water and preserves embodied energy (Baker-Brown, 2017). This particular line of development concentrates on the management and reuse of existing capital in

DOI: 10.4018/978-1-7998-1886-1.ch005

a loop fashion where the life of products and their components are extended, resource consumption is reduced, and waste is considered a new input material for manufacturing processes.

In some countries the designers are already changing their attitude. For example, to support architectural firms, the Royal Institute of Dutch Architects (BNA, 2018) has written the Manifesto Circular Architecture together with its members Royal Institute. Document contains five design principles that might seem quite obvious, but stimulate a different path to design through offering tools for both the design itself and for the necessary discussions with all stakeholders. Their contents is shown below:

1. A circular business model is the starting point for circular architecture.
2. Nature is a source of inspiration and a textbook example of circularity.
3. A structure is adaptable and flexible throughout its life.
4. A building including its components is easy to (dis)assemble and construct.
5. The building materials are of high quality, non-toxic and easily reusable.

In fact, this Manifesto confirms the main paths of changing attitude which have been discussed in this book so far.

Circular economy principles could offer the construction sector a transformative possibility by considering concepts such as modular design, prefabricated and off-site construction, design for disassembly, materials recycling and designing out waste (From linear to circular, 2018). When, and if included in the construction industry, circular urban and architectonic design as well as construction processes will lead to the modification of contemporary standard design and building process evolving into interdisciplinary flexible co-operation, integrating experts from various disciples. Longer pre-design phase should include more advanced research, specialist consultations, building material tests and experiments. This phase should also include interdisciplinary design team and input from future contractors. Adequate use of secondary materials in building industry requires often detailed and even more often expert estimation of their technical data, level of preservation, type of primary use, durability, chemical content, toxic and pollution levels, environmental impact and estimation of possible imperfections. Interdisciplinary team work during concept design stage allows to define the level of the recycled materials' potential use in new functions – defines optimum sourcing procedures, remanufacturing processes, mounting and finishing procedures, as well as estimation of costs and time (Addis, 2006). The choice of secondary materials includes several technical, aesthetic, environmental, economic and social criteria, which in turn may become a very cost consuming design process. This approach proves the need for general specialist research proving conformity with existing standards, often acquiring certification and permissions prior to preparation of building materials specification. In contradiction to a typical process where the obligations are part of the manufacturer's duty, in case of recycled materials presently it is the designer who has to receive adequate permissions (Addis 2006). Flexible cost calculation and time schedule should include unpredictability of the secondary resource market, their limited accessibility, the requirement to seek sources and non-existence of standard methods and procedures. The main environmental aims initially formulated as a planned share of recycled or re-used building materials have to be presented in the initial stages of the project and explain in more detail in the design condition and materials specification. Attainment of the environmental priorities must be monitored by a team of qualified experts during every phase of design and construction. Any design set forth on the condition that after the end of life building elements may be re-used, also differs from a standard design procedure. Similarly, as when using recycled materials, pre-design phase has to include a wide variety of interdisciplinary expert opinions. Sometimes research checking the case best form the building which later will allow for re-use of building materials should be included in design procedures. Expert opinions from specialists in selective dismantling and recycling processes of building waste can be act as support knowledge to distinguish which materials can undergo recycling process and which processes are considered as simple and financially viable. These experts can also give information on design of joints for easy dismantling process after the end of building's life. There is also a requirement to develop new dismantling methods and building techniques, as many of the modern materials such as reinforced concrete and plastic components are "melted" into a single material at very intricate levels. Additionally, interdisciplinary co-operation between architects, structural engineers and other technical specialists should start within the early documentation phase and must be continued throughout the construction process. This special knowledge can be later used to prepare tenant and maintenance manuals, as well as dismantling manuals. Specification must include foreseen future waste flows and most efficient procedures for reclamation of building materials. Both, design with the use of recycled building materials and for the future dismantling of the designed building are the base for circular economy implementation in building industry. Circular design does not only mean a longer life of used products achieved through durability, modular solutions, maintenance or repair. More care should be given towards the user requirements covering design of circular buildings, products and services, as well as complexity of the process due to the need to adapt products to more than just a single life cycle and more than just a single user (Lofthouse, Prendeville, 2018). A variety of users furthermore means that provided product should be neutral without emotional ties of the user (Tunn et al. 2018). Products and buildings should be designed in a way which will allow to maintain the case best value of the materials during long periods of existence and the lowest environmental, economic and social impact possible. Such approach was presented in the Cranfield University Model to Extend the Life of Materials – MELM (Encinas-Oropesa et al 2018), which consists of five parts. Material flows – waste and waste valorisation, re-manufacture of materials, prototypes, integration of technical and social factors. In order to recognise parameters and performance of materials this model uses data from standard material tests and advanced technologies (i.e. uses gravimetric analysis, microscope scan, Roentgen radiation). Additionally, to monitor material flows and their environmental impacts, MELM uses LCA method a Circularity Indicator (CI). This tool may be used for analysis and efficient circular use of materials and their components. One of the most know circular building examples of Super Circular Estate in Dutch Parstad designed by ZUYD University of Applied Sciences and CIRCL Pavilion in Amsterdam, a living laboratory experimenting with innovative solutions. Building Circularity Index (BCI) - a calculation method to indicate circularity of buildings based on disassembly and waste scenario of materials and products (https://albaconcepts.nl/building-circularity-index/), and Cirdax – online application from Re Use Materials to register materials of existing buildings and manage re-use scenarios (www.reusematerials.nl/over-ons), may also prove to be useful applications.

The choice of building materials and structure is one the most important areas within the design process conducted by architects, interior designers, engineering architects and urban planners. This choice has impact both on the aesthetic, flexibility, adaptability durability and structural issues. Presently it should also include circularity, even though this particular area has been within the reach, since the beginnings of our Civilization, even if our contemporary circularity is based much more on the abundance than scarcity. Our profession locates us within the pre-phase of the building industry, and our choices may prove to be of major benefit both to the environment, as well as to the creation of a waste-free carbon-neutral society.

According to prof. Walter R. Stahel (Baker-Brown, 2017), contemporary building industry faces a triple challenge of:

- Developing efficient and a waste-free construction method, characterised by the ability for later re-use;
- Design of flexible and adaptable buildings, with efficient resource consumption;
- Developing of methods enabling deconstruction of buildings and infrastructure while maintaining the highest values.

To this list, we can also add two additional challenges:

- Introduction of education for the architects, interior designers, engineering architects and urban planners, so that their choice is guided not only by efficiency and environmentally friendly solutions, but will include aesthetic choices;
- Introduction of education for a wide scope of societies to make them understand what is the aim of such transformation and what are the responsibilities at the users' end.

Possibly this book may act as a guide to the students of Architecture and Urban Planning explaining the nature of circularity in design making process, including management of issues which should be set forth according to the particular site's parameters and development possibilities. Use of feedback and understanding of efficiency and scale of used solutions. Legal issues as part of the framework where circularity is considered will be discussed. Introduction of different scales allowing to match the outcomes of linear and non-linear approach to the city making choices. A framework road map for circular loop design which will include issues which should be checked, including: circularity in building materials, structural waste and approach to economic losses, efficient energy, social approach, changing cultural approaches and negative environmental impacts, urban mobility system and issues on urban bio-economy.

It has also been noticed that user perspective has been overlooked in the circular building approach (Geldermans, 2018). A number of publications omit the issue of flexibility and possibility to change the internal configurations including function change. There are multiple approaches, one of the ideas being that the structural and non-structural decisions may be undertaken by different stakeholders (both investors and users). In this type of organisation, the process could start on the smallest unit – a room – and be scaled to apartments, buildings, cities or regions – it could be seen as an organism. Unfortunately, presently most of the dwellings are designed for a single type of occupant expectations, so diversity is allocated to the interior of people's homes and is subject to changes due to increase or decrease in family size or new accessibility requirements. This issue is also undergoing research with the post-occupancy evaluation (POE), also used in some certification programs i.e. Home Quality Mark (HQM used within BREEAM procedures www.breeam.com) or being part of the WELL Certificate (www.wellcertified.com). Therefore, circularity issues should also include such aspects as the possibility to alter interior space-plan, with the ability to change the non-structural components according with the user expectations, this in turn may help in description for recycling paths for the surplus materials which otherwise would become waste. Such circular and flexible building concepts (Circ-Flex) should be of user and economic benefits such as: ease of maintenance, redistribution, remanufacturing, recycling, facilitating bio-cascades and bio-feedstock (Geldermans, 2018).

The idea of circular design has become a subject of several Master degree thesis. Two exemplary case studies are presented in the Appendix. Both initiated from their authors' interest in circular use of materials. The description has been authored by the students of Architecture who have been awarded a Master Degree in Architecture based on their design analysis. Author of this book was a reviewer for the first presented Diploma, and a Promoter for the second one.

Appendix no 1 focuses on the implementation of sustainable solutions in practice, but on an on an early stage of design process – a Concept. Analysis explores optimisation of a building's life cycle impact within the chosen site. Research conducted in a BIM-based LCA includes the comparison of building material alternatives and is focused on the best case solution. Dissertation also includes predicted life cycle costs, energy use, carbon footprint, net use of potable water resources, and embodied energy calculations. It efficiently follows the process down to presentation of a building which could be constructed with the use of above tools and solutions.

Appendix no 2 presents a more futuristic approach dealing with human space travels and construction of scientific bases on different planets. This development is a subject of various discussions, but as yet circular aspects have not been touched. The thesis indicates that the circular aspects should be applied simultaneously to the Martian and earthly contexts, providing benefits for both planets and their inhabitants. The dissertation follows a discussion on the process of Mars exploration and the interplanetary transport posing a major challenge. The main argument is that use of the in-situ resource and automatic, additive manufacturing of the habitat can minimize the required transport costs. According the student (and his assumptions were scientifically based) most of the resources is available directly on the Martian surface in the form of regolith and atmosphere. Outside the construction materials, the rocket fuel (methane and oxygen) for the transport system can additionally be easily synthesized on Mars. Required atmospheric CO_2 and water are present just below the surface (up to 10% of soil mass).

This approach was also used in various design studios conducted at the Faculty of Architecture Warsaw University of Technology. Proposed general curricula covered implementation of practical professional use of sustainable and circular development solutions directly influencing urban and building scale planning, design of architectonic forms as individual buildings located existing city context. It compiled knowledge which can be used at each and every level of design and construction process. Emphasis was placed to treat the climatic parameters as a valuable asset from the very start of design. This also called for a wider implementation of knowledge concerned with natural environment and its preservation and loop solutions. Two important areas were distinguished:

- The need to integrate sustainable and circular development of societies and urbanization procedures, specifically within the design process, organization and planning of the building investment process, and facility management as well as retrofit of already existing buildings
- The need to introduce and implement a holistic multilevel understanding of the urban space, including harmonious interactions between manmade and natural environment within curricula of architectural education

The scope of knowledge required, concerned an input of sustainable and circular solutions taught during lectures within the curricula of legislation, economy and organization of building investment process – from the concept level, through construction, use, modernization, and building's life cycle, as well as possibility to re-use some of the structural and building elements for recycling. Analysis were shown and prepared proving that orientation and shape had a dramatic influence on a building's energy needs. Students were asked to decide what would be their choice when allowing to take advantage of daylighting and natural ventilation, and perceive how that would made an impact on the orientation and shape of the final building's volume and detail. They were asked whether use patterns and the layout of the space offered any opportunities to determine the energy demands. Natural light used in multiple spaces, possibility to reuse heat emitted in some areas of the building in other areas of the building. This was an interesting and widely pursued area, and one of the groups developed an option where surplus heat energy was transported from a nearby metro station to their designed building. Students started off with a Client meeting, during which the Client indicated the site and presented the idea as to the function and program of the building. Following seminar was dedicated to a trip – and individual check of the site and its urban context. They met regularly with academic staff and every second week with consultants representing urban planners and environmental engineers (alternative energy consultants, environmental consultant). They also met regularly (once every two months) with a client team, to present and discuss their design.

There are two projects which should be mentioned. The first one took place in 2016 and was prepared at the request of suburban Warsaw area called Jablonna. In this particular case representatives of the local authorities wanted to acquire a set of methodological tools allowing to design and later construct a Local Point for Selective Collection of Communal Waste. Existing structure was divided into three distinct functions: office, conference area and waste selection area used together with seven external containers. Program also included children education path, a flea market and a small parking site for the clients. Prepared semester designs are original and often use innovative architectural solutions. Both natural and re-use materials were used very often. Students also tried to change the social perception of waste, through re-use as a good standard building material. Approach included better management of building materials wastefulness within design process which included Upcycling and preparation of a local Harvest Map, promotion of fab labs – small workshops offering personal digital fabrication. In one of the cases the main idea was design of façade with the use of re-used materials collected from local metal scrap yards, this approach included a location map. In this case the costs are limited to transport only, and purchase of sub-mounting system. Concept proposeed construction of façade from square and rectangular pipes (50x60mm up to 50 x 150mm). Example is shown on Figure 1

Unfortunately, this approach proved to be inacceptable by the local authorities, as the circular choices were not understood as the best case ones. More, as the ones that required more work where building materials

Figure 1. View of the proposed waste unit
Source: (students: K. Sobiecki A. Sztyber, Economy and Management of Building Process, semester 8, Bachelor studies, 2013/2014, Faculty of Architecture WUT)

were concerned and even more work if the building was to be designed and constructed.

The second project concerned cooperation with Warsaw Copernicus Science Centre, with given site is located in the vicinity of exiting building where underground parking lot was already in place. The issue was how to locate the new volume on existing structural grid and supporting columns in order to reuse and repurpose existing structural solutions, and check whether the existing structure had the capacity to carry the newly designed loads.

Figure 2. Main façade of the existing Copernicus Science Center and the newly designed building
Source: (students: K.Plawny, K. Kierznowska, Design Studio semester 1, Master studies, 2014/2015, Faculty of Architecture WUT)

Three distinct solutions were perceived:

- Solution no 1 (example shown Figure 2) – students decided to strengthen existing structure and remain within the perimeter of existing GFA of the ground parking lot. This resulted in a series if compact – cube-like volumes. Nevertheless each of the proposals has individual aesthetic features.
- Solution no 2 (example shown on Figure 3) – students decided to design new structural elements outside the perimeter of existing walls, but design the GFA within the underground parking lot. This resulted in rectangular buildings, often terraced with many green roof spaces.

Figure 3. Main façade of the existing Copernicus Science Center and the newly designed building
Source: (students: P.Boruch, K.Katerla, K. Klik, Design Studio semester 1, Master studies, 2014/2015, Faculty of Architecture WUT)

Figure 4. Main façade of the existing Copernicus Science Center and the newly designed building
Source: (students: J.Michajlow, A.Szajda, A.Swiderska, Design Studio semester 1, Master studies, 2014/2015, Faculty of Architecture WUT)

Solution no 3 (example shown on Figure 4) – students decided to design new structure altogether. This resulted in choice of design where buildings were either totally submerged underground (with the use of existing structural elements) or large volumes of terraced green roofs which made the buildings seem like part of the landscape

All of the students were asked to use natural ventilation system, adequate daylighting strategies, and alternative energy sources with emphasis in circular energy solutions. The interesting issue is that, most of proposals actually were prepared within the budged proposed by the Client. The compact cases had more floors and additional financing was required for the strengthening of existing structure, whereas lower rectangular terraced buildings used less expensive structural solutions, but more expensive green roof solutions. The only project that was out of the budget – was 90% submerged, in return, feasibility study showed that exploitation costs could be cheaper by 20% in comparison to more traditional solutions, as part of heat gains from local metro could be reused for heating purposes. Also, due to underground solution, outside complying with effective flow of existing building materials, this complex appeared to have more stable internal parameters than any of the other proposals. One of the proposed solutions was used to prepare a client's brief for announced competition and included strong emphasis on re-use of existing structural members and other building materials.

October 2019 brought a workshop for the students of Architecture entitled Circular Design for Architects and Urban Planners. It formed a part of the Warsaw Circular Week: 7-13th October 2019 and was prepared with the support of arch. M. Rubbens from the Dutch CEPEZED architect studio (sponsored by Dutch enterprises including Embassy of Netherlands in Poland). This workshop aimed to create an awareness in the students that when approaching a new site, they should include regenerative pathway of design both on urban and architectural scale. This means that within the design process they were required to:

- Sustain and preserve what is already on the site - maintain, repair and upgrade resources in use to maximize their lifetime.
- Use waste as resource – utilize waste streams and recover waste for reuse and recycling; prioritize regenerative resources.
- Design for the future - adapt a systematic perspective during the design process to employ the right materials for appropriate lifetime and extended future use.

- Work in a team to create a joint value – internally and with organizations to create shared values.

Chosen site was part of industrial Warsaw, with abundance of small industrial enterprises which were become part of the design. Students were divided into two working groups: one dealing with materials flows and creation of shared values on the urban level; the second dealing with design of a multipurpose building on the given site. Workshop had its share of thematic lectures dealing with circularity. The urban planning students focused on the function related to the automotive industry and proposed to:

- Create a system structure for the circulation of goods within the district.
- Introduce additional complementary functions for the circulation - car dealership, "restore" shop with unused materials from factories, place for dismantling cars, stores with used parts, logistics and laboratory centre, processing of car tires into granules, processing of plastics..
- Creation of links between existing and planned functions

Where design of buildings was concerned main focus was on the process of designing of sustainable building starting at the planning stage of a building and continued throughout its life to its eventual deconstruction and recycling of resources to reduce the waste stream associated with demolition. For this reason, the students felt that the architects should be responsible for the future and consider each building planning detail just at the beginning of designing. Interdisciplinary thinking ensured diversity, which made our discipline adaptive. Most of the participants concentrated on various types of shipping containers as on-site building resource with capacity to change and future relocation.

Close contact with the perception and approach of students, who wish to deepen their knowledge on the possible implementation of circularity and loop economy in building design shows that such approach is possible. It also enhances the fact that this is an area still under research, and a lot of analysis have to be made before circular economy will become a path which can be followed and become an indispensable element of future.

CLOSING REMARKS

It should be remembered that in contrast with the main economic path, the circular economy is still being developed. Optimal design of closed loops when based on sustainable conditions – not just those concerned with construction industry - should result in radical reduction of waste and increased economic values. The main concern is whether these circular solutions can be realised in practise – regardless of very positive scientific approach. For example, Friege and De Man (De Man and Friege, 2016) make an argument that waste as such cannot be accepted as a non-degraded material (either food or other resource), following that they state that we need additional energy to reprocess the end products and therefore high level of energy has to be used in order to achieve a waste free economy. The other issue they raise, is that our knowledge whether such solution will sustain harmless effects, is only emerging and since throughout the development of human civilisation production always resulted in abundance of industrial wastes – this case might prove to be the same. Hence, they also argue that the assumption where circular solutions lead to sustainable outcomes is erroneous. Friege and De Man (De Man and Friege, 2016) also argue that sustainable resource management will start from an overall optimal concept scoping such issues as waste policy, resource management, efficient energy and climate protection. Hence, their idea that EU policy should focus on instruments that aim to reduce the consumption level of materials and increase materials and energy recovery may be followed especially in construction industry where waste has always been abundant. They also point out that a new set of rules for design of specific products would support "design for recycling" process in the architectonic design and urban planning processes. This statement is further supported by other scientists (Spoerri et al. 2009), (Peiró et al. 2019), who try to illustrate how circular economy strategies can be implemented by European product policies.

Others such as Korhonen and Seppälä (Korhonen and Seppälä, 2018) discuss limits to the circular economy, naming thermodynamics and entropy issues; spatial and temporal system boundary issues; limits posed by physical economic growth; path dependencies and lock-ins; multileveled organisational issues; and possibility of physical flows. All of those areas also create possibilities for further research for many different disciplines.

There are other arguments at hand as well. For example Esposito, Tse and Soufani (Esposito et al. 2018) point out that waste should not exist at all when and if a product is designed in an appropriate biological and technical cycle and may be used several times with lower embedded energy. They also show, that this application is possible in construction industry, with high volumes of waste due not only to inefficient construction procedures, but also wrong decisions undertaken in the concept design stage. Another area which has been marked for further research area is manufacture process of building materials and their transport. One of the routes which may be followed is 3D printing as more sustainable than traditional production processes, but still having a potential to prove negative outcomes (Ruby, 2010), (Unruh, 2018).

This new approach to circularity also takes the ideas from metabolism (Andenberg, 1998); (Niza et al. 2009), (Chong et al. 2010).

(, which in itself is not new, but more focussed on product and service design, engineering and business models). These models are complex and flexible; emphasise participation, connections and consequences Fig.29 (Ellen Mac Arthur Foundation, 2016). Loop economy is considered in aspect as regenerative, where waste (regardless whether technical or biological) is considered as further input, also pointing at need to shift to the use of renewable components. A lot of attention is posed on the level of available information for all stakeholders, as only such approach will allows for sufficient diversity and strength. Additionally, the design, construction and harvesting process have to be closely interlinked and stakeholders are expected to cooperate and form chains managed according to guidelines involving all interested parties. This general definition has its counterpart in design and construction but requires a different way of the processes understanding, starting from the site with the main assumption that the new structure

will be one day dismantled and therefore circular choices have to be the initial ones together with the sustainable issues. Analysis concerning reuse of products and materials from the vicinity or site itself is part of the approach and set as prioritization. Building is perceived as a collection of layers – functions, elements and products – each with a different lifespan.

According to Circle Economy and Netherland Circulair! Report, calculation of the net present values and climate and energy systems demonstrate value from circularity in separate layers with higher residual value of elements, products and materials. Hence it makes more sense to analyse layers not whole buildings (Fischer, 2019.). The most durable layer is the structure – circa 100 years, envelope – circa 80 years and services circa 40 years. It should be remembered that the services might be expected to be changed earlier according to the user preferences and expectations.

Harald Friedl from CEO, circle Economy states that the best solution is modular structure, where everything can be replaced, remanufactured or

Figure 5. Schematic of Regenerative Circular Economy
Source: Circular Economy and curriculum development in higher education. Briefing notes, support & illustrative resources. (Ellen Macarthur Foundation Higher Education Programme, 2016).

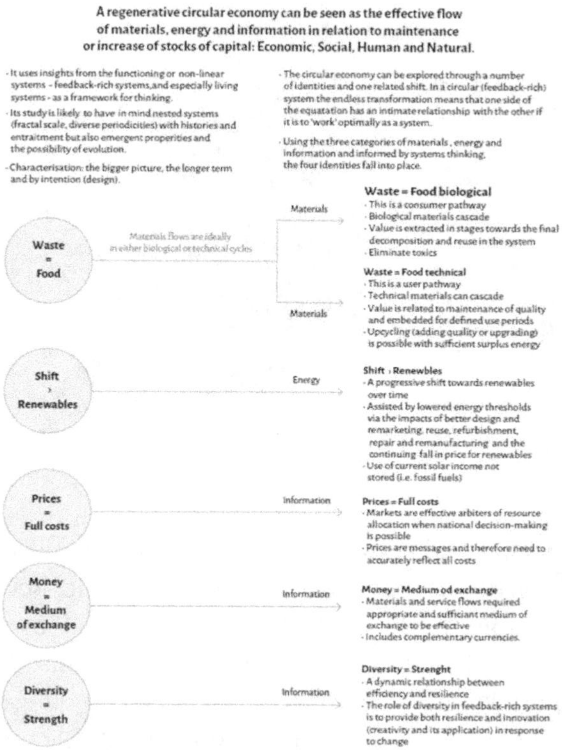

recycled. There is also a notion of a Value Hill – Fig. 30 (Achtenberg et al 2016), which originally was a diagram showing different strategies to retain product's value. This diagram was adapted by the Circle Economy for circular construction processes and such has been used below.

The Value Hill concept when applied within construction sector initially deals with the re-use of elements – which is the present attitude with the lowest value-strategy, then re-use products and in the last phase - re-use of materials. This diagram shows the high energy consumption for new building materials and reflects preference to maintain the value of in-built products and elements. In order to follow this path, there is a need to design in a way which will allow adaptability beyond original intent. Design flexibility may be included on various levels and different design stages. The most widely knows idea is an open plan. Additional issue is transparency in used materials and products, as all data must be collected and exchanged. In order to re-

Figure 6. The Value Hill in construction sector
Source: (Fischer et al. 2019)

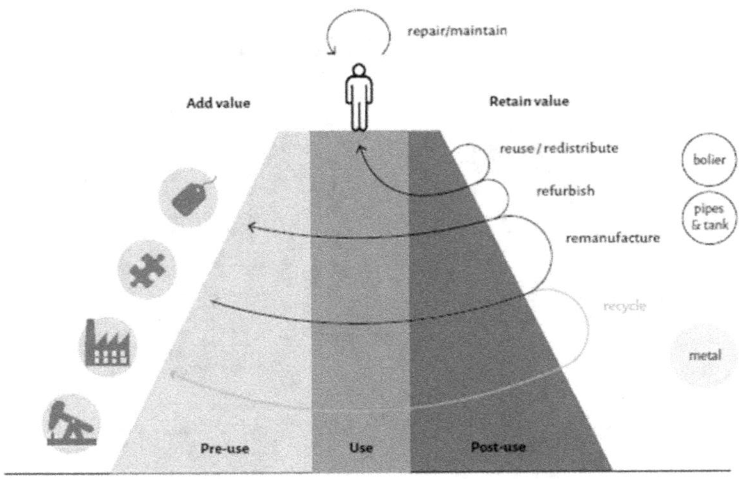

use building materials stakeholders have to know how these were originally mounted. Possibly this standardization process could become part of the Building Information Models (BIM) and supported by Building Smart organisations where their Industry Foundation Classes (IFC) model uses BIM, but also includes circular concepts.

EEA surveys indicate that circular economy in Europe is very much still in its infancy, and possible progress requires further investment especially where available data is concerned – this covers both production and consumption phase of products LCA often not available on the market (EEA, 2019c). Much emphasis is show towards the fact that circular concept should be understood as part of a broader shift in research and policy. In EU context it may be noted as appearance of strategic frameworks and action plans. These approach is based on an understanding that the ecosystems sustaining our society are finite. Transformation cannot be achieved through simple or even more advanced technological fixes and applied efficiency measures. This particular transition also requires new business models, social approaches, lifestyles and further on prove to be a source for new jobs. When considering not just European, but the world market there is a need for new knowledge in such areas as raw material cycles, position of producers, designers, manufacturers, consumers and recycling companies. There is a requirement for innovative technologies and data analysis which may allow to trace the flow of materials much better than presently.

For designers and urban planners the most important area to be pursued should be reuse, repair, remanufacture and refurbishment, which so far has not as yet been given sufficient attention in comparison to recycling and waste disposal. Some of the advantages are obvious as choice of more durable materials limits the use of new natural resources and lowers impacts associated with production of new materials. Circular products design has to be supported by a strong governance system and a monitoring system, as there is little value in changed design if there is no infrastructure for harvesting on the large scale urban planning model. Some countries have inserted those issues into their economic policy i.e. Flanders (Belgium) has mentioned implementation of a circular economy map with targets in 2021-30 plan linking circular and climate policies (EEA, 2019c).

CONCLUSION

Above analysis present a theoretical, under research approach. Unfortunately, for Architects and Urban Planners a straightforward approach does not exist as yet. Those, who follow the road of circularity try to establish a foothold through design and construction volumes and solutions to be monitored and repeated at a later stage. This is complicated research area for all stakeholders involved in construction sector with many question marks and unknown developments, and hopefully this book has given at least a few ideas as to which paths and solutions may be followed. It should be noted that this book should be mainly used by the students of Architecture and Urban Planners, which in many areas still lacks the interdisciplinary approach. There is also a need to experiment with new software allowing for parametric design. In many countries interdisciplinary design approach is still in its cradle especially where urban planning is concerned, and possibly a different approach may be formulated when circularity will become a part of the urban "draft board", working together with geographers, sociologist, and technical engineers.

REFERENCES

Achterberg, E., Hinfelaar, J., & Bocken, N. (2016). *Master Circular Business with the Value Hill* (White Paper). Financing Circular Business. Retrieved from www.circle-economy.com/financing-circular-business

Addis, B. (2006). *Building with reclaimed Components and Materials*. Earthscan. doi:10.4324/0781849770637

Andenberg, S. (1998). Industrial metabolism and the linkages between economics, ethics and the environment. Ecological Economics, 24, 311–320.

Baker-Brown, D. (2017). *The Re-use Atlas. A designer's guide towards a circular economy*. RIBA Publishing.

BNA. (2018). *Royal Institute of Dutch Architects. Manifesto Circular Architecture*. Retrieved from https://www.dutcharchitects.org/we-are-going-circular/

Bundesministerium für Umwelt (BMU). (2002). *Technical Instructions on Air Quality Control*. TA Luft.

Chong, W., & Hermreck, C. (2010). Understanding transportation energy and technical metabolism of construction waste recycling. Resources, Conservation and Recycling, 54, 579–590. doi:10.1016/j.resconrec.2009.10.015

De Man, R., & Friege, H. (2016). Circular Economy: European Policy on shaky ground. *Waste Management & Research*, *34*(2), 93–95. doi:10.1177/0734242X15626015 PMID:26759489

EEA. (2019c). *Paving the way for circular economy: insights on status and potential*. EEA Report no 11/2019, EEA. Retrieved from www.eea.europa.eu

Ellen MacArthur Foundation. (2016). *Intelligent Assets: Unlocking the Circular Economy Potential*. Ellen MacArthur Foundation. Retrieved from http://www.ellenmacarthurfoundation.org/assets/downloads/publications/EllenMacArthurFoundation_Intelligent_Assets_080216.pdf

Encinas-Oropesa, A., Moreno, M., Chamley, F., & Simms, N. (2018). A Model to Extend the Life of Materials, Circular Economy disruptions, past, present and future. In *International Symposium Abstracts 2018*. Ellen McArthur Foundation. Retrieved from https://www.ellenmacarthurfoundation.org/assets/downloads/Circular-Economy-Symposium-Extracts-June-2018.pdf

Fischer, A. (2019). *Building Value. A pathway to circular construction finance.* A Report by Circle Economy and Netherland Circulair! with the support of the Community of Practice. Retrieved from www.circle.economy.com

From linear to circular Economy. Experiences from Denmark and New York on closing the loop through partnerships and circular business models. (2018). *State of Green*. Retrieved from www.stateofgreen.com/en/publications

Geldermans, B. (2018). Circular & Flexible Building: For whom?gelder A user perspective. In P. Luscuere (Ed.), *Circulariteit. Op weg naar 2050?* Tu Delft Open.

Korhonen, J., & Seppälä, J. (2018). Circular Economy: The Concept and its Limitations. *Ecological Economics*, *143*, 37–46.

Lofthouse, V., & Prendeville, S. (2018). User centric circular solutions and implications for design. In *Circular Economy disruptions, past, present and future, International Symposium Abstracts 2018*. Ellen McArthur Foundation. Retrieved from https://www.ellenmacarthurfoundation.org/assets/downloads/Circular-Economy-Symposium-Extracts-June-2018.pdf

Niza, S., Rosado, L., & Ferrao, P. (2009). Urban Metabolism Methodological Advances in Urban Material Flow Accounting Based on the Lisbon Case Study. *Journal of Industrial Ecology, 13*(3), 384-405. Retrieved from https://www.researchgate.net/publication/263564766_Urban_Metabolism_Methodological_Advances_in_Urban_Material_Flow_Accounting_Based_on_the_Lisbon_Case_Study

Peiró, L., T., Polverini, D., Ardente, F., & Mathieux, F. (2019). Advances towards circular economy policies in the EU: The new Ecodesign regulation of enterprise servers. *Resources, Conservation & Recycling*.

Ruby, A. I. (Ed.). (2010). *Re-inventing construction*. Berlin, Germany: Ruby Press.

Spoerri, A., Lang, D., Binder, C., & Scholz, R. (2009). Expert-based scenarios for strategic waste and resource management planning—C&D waste recycling in the Canton of Zurich, Switzerland. *Resources, Conservation and Recycling, 53*, 592–600. Retrieved from https://ethz.ch/content/dam/ethz/special-interest/baug/ifu/eco-systems-design-dam/documents/lectures/2015/master/prospective-environmental-asses/readings/ifu-esd-msc-PEA-Spoerri_et_al_2009_C_D_Waste_FSA_2013.pdf

Tunn, V., Schoormans, T., Hende, E., & Bocken, N. (2018). Non-personal Design: Lowering Acceptance Barriers of product-Service Systems. In *Circular Economy disruptions, past, present and future, International Symposium Abstracts 2018*. Ellen McArthur Foundation. Retrieved from https://www.ellenmacarthurfoundation.org/assets/downloads/Circular-Economy-Symposium-Extracts-June-2018.pdf

Unruh, G. (2018). Circular Economy, 3D Printing in the Biosphere Rules. *California Management Review*, *60*(3), 95–111. doi:10.1177/0008125618759684

KEY TERMS AND DEFINITIONS

Autodesk Revit: Intelligent model-based process used to plan, design, construct, and manage buildings and infrastructure. Revit supports a multidiscipline collaborative design process, is often used as one of support open programs in BIM environment.

Case Study: Architecture Laboratory Building (ALB)

Architecture Laboratory Building (ALB). *Building's life cycle issues based on the design of a research institute next to Pole Mokotowskie Park in Warsaw, a Master degree thesis, promoter: Marcin Goncikowski, PhD. Arch. Reviewer: Elzbieta Rynska. Presented at the Faculty of Architecture, Warsaw University of Technology.*

ABSTRACT

As architects, we design to develop and improve our surrounding environment. Meanwhile, the way we put our visions into shape today guides us to quite contrary results. Current impact of building industry brings unbearable burden to the environment, and puts it to a risk of total eradication. This paradox was the beginning to the PhD. thesis: Architecture Laboratory. In this work it was investigated how architects can shape their visions in the world challenged by climate change, working under the imperative of sustainable development.

The first part of this paper presents the object of study - the building of Architecture Laboratory, whilst the second focuses on the implementation of sustainable solutions in practice, specifically in a BIM-based LCA. The project explores ways to optimise a building's life cycle impact within the use of a native design environment. Analysis is delivered on an early stage of design process, based on a conceptual model combined with LCA data. Exploration includes the comparison of different building material alternatives, amongst which the most sustainable composition is sought. The scope of work includes such analysis as predicted life cycle costs, energy use intensity, carbon footprint, net use of fresh water, embodied energy amount and more.

Keywords: design for sustainability, life cycle analysis, life cycle cost, energy analysis, BIM, Dynamo, Revit, computational design, design automation, EPD.

INTRODUCTION

Function

Architecture Laboratory building is a new unit of the Institute of Building Technology in Warsaw, Poland. It's function is dedicated to research in recycling and reuse of materials from existing buildings - waste from renovation and demolition. This unit was designed specifically to support the implementation of circular economy in practice, addressing the problem of waste in C&D at its basis (EC, 2016)[1].

Location

The project is located in Warsaw, next to Pole Mokotowskie Park, in a short distance to Warsaw University of Technology campus and other units of the Institute of Building Technology. This location was chosen due to possible interactions with visiting people, enabling it to function as an educational institution that attracts attention and disseminates information about circular economy.

Architecture

The building has three levels – basement, ground floor and first floor, with vast majority of functions located to the ground floor. It spreads horizontally, with one solid dominant - cube of the main research laboratory. South part of the building is dedicated to exploration and is open for all users. Workshop area for seasonal research activities is located centrally. Remaining part is restricted research area, with large workshop laboratory and small offices surrounding it.

Building's architecture was determined in conceptual boundaries, providing such information as form, orientation, function and daily operation scheme. The BIM model included the main vision of this project, which only later was to be augmented with LCA data.

One of the greatest challenges in this project was preparation for the use of various materials. As a compromise, a structural module of 6.6 meters was adopted. This enabled a calculation of a valid option for both materials such as steel and concrete, as well as compounded mixture of timber particles and rammed earth.

Appendix 1

Figure 1. Exterior view from east side, with entrance to the green roof
Source: (Master thesis presented 2018)

Figure 2. Interior view of main research laboratory
Source: (Master thesis presented 2018)

Figure 3. Exterior view from Pole Mokotowskie Park
Source: (Master thesis presented 2018)

Figure 4 Ground floor plan (AL Building).

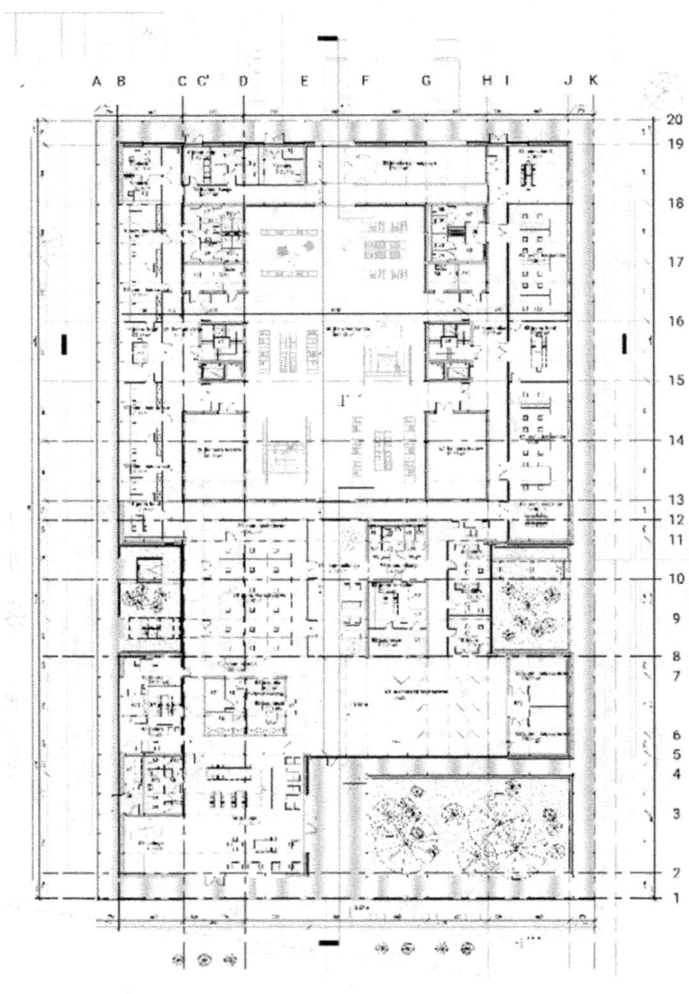

Figure 5. Longitudinal section of the AL Building.

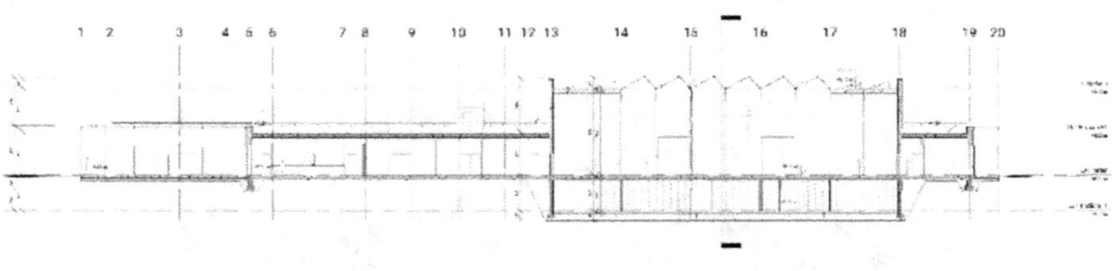

Figure 6. East elevation view of the AL Building.

Examination Methods

The main aim for this study was to analyse a building's environmental impact and to recognise the best material solutions for given building concept. In addition to the environmental impact analysis, also building's life cycle cost (LCC) and predicted energy use intensity (EUI) were calculated. This way, a set of analysis provided a wide spectrum of data needed to assess alternatives for a competitive market.

Object of Study (Information Model)

The basis for this study is a conceptual building model prepared in *Autodesk Revit 2019*. The most important task in model development was preparation for flexible data assessment. All elements of the model were defined in a way that would allow smooth collection of quantitative data. Walls, floors, columns and beams were represented with geometry containing information about their areas and lengths, to be later substituted with respective values of target compound elements (layered floors and walls) and profiles (columns, beams), using computational design tools as described further in "LCA assessment" part.

Target elements for each material alternative are included in the model, as integrated database of building components. Each wall, floor or column has defined material layers, and each of its materials has defined attributes. For this study, Revit material library was extended with EPD data, applying to them additional parameter values as declared in EN 15804 (PKN, 2014)[2].

The last set of information included in the model was intended for energy use analysis. Model was supplied with function, operational building scheme, predicted number of users and estimation of installed HVAC systems, included in built-in *Autodesk Revit* and *Insight* tools.

Figure 7. BIM model – 3D section with marked building elements analysed in LCA study

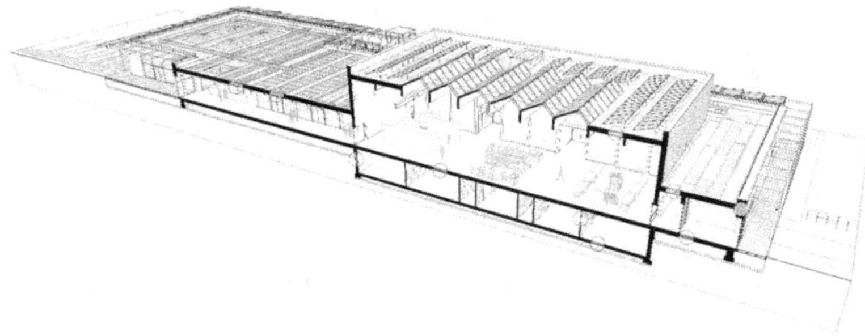

Material Alternatives (AL Building)

In the scope of project it was decided to asses five alternative material compositions, which differ in core structure and infilling materials, as presented below. Material choices were based on the observation of popular materials used regionally in Poland, as well as the research on possible alternatives to these materials.

Building elements varied in the materials and components used for walls, floors, roofs, columns, beams and stairs. These alternatives were applied only to the building's superstructure. Regarding the underground, the same type of structure and building materials were used - concrete structure with calcium silicate units as infilling walls, concrete foundations, columns, and

Figure 8a. Definitions of the icons used in the table of building elements for analysis (please see below)

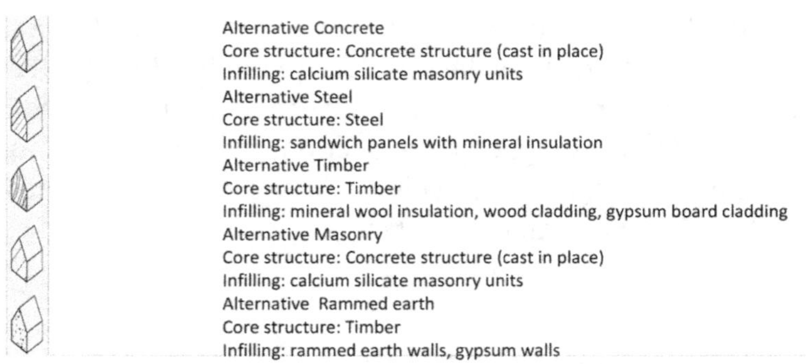

Figure 8b. Table of building elements for analysis. Columns represent each component type (ie. exterior walls, floors), while in rows represent each material alternative (AL Building)

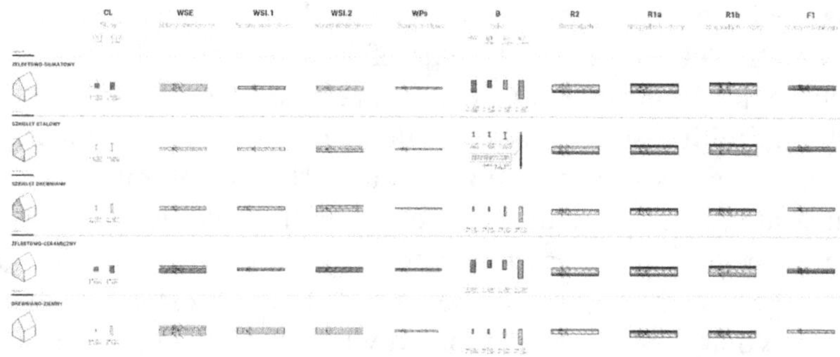

floors. Another common element was glazing, calculated as the same type in each design option.

LCA Assessment

Accordingly to ISO 14040 (PKN, 2009)[3] LCA assessment was delivered based on EPD data. For this project IBU (IBU, 2018)[4] and ITB (ITB, 2018)[5] EPD's database were used, which provided values for A1-A3 modules (product stage) impact calculation – the impacts embodied in the materials used in construction. All information about crucial emissions, waste, water and energy use was added directly to the BIM model, into the properties of building materials. Calculation was based on EPD data per cubic meter. Due to that fact, special attention was paid to the consistency of units.

Next step was a collection of quantitative data from the model. The target was to estimate volume of materials used in each design alternative, and calculate their emissions and resource use respectively. Usually in a more developed design stage with better model accuracy (such as LOD 300 and higher) one could use regular quantity-take off for this matter. However, for the purpose of a concept design it was decided to explore possible optimisation method allowing a more flexible assessment. An attempt was taken to create a model that would enable comparison of various material alternatives without the necessity to prepare a detailed, detached model for each design option. Computer software supported design enabled avoidance of tedious work.

Using Dynamo for Revit, a dedicated algorithm was developed, it helped to collect quantitative data from the conceptual model and link them to the

LCA data library. The algorithm collects basic quantities from the model and multiplies them by target material parameters - for each building element, in every design option. The core of this algorithm was to estimate volume of the materials used in each design alternative, which was done based on the basic model geometry definition. Another part of the algorithm definition combined these values with appropriate LCA data parameters from the model. The last part of the definition summarised results and organised them according to building element types - separately for walls, floors, columns, etc.

To illustrate it with an example, to estimate the volume of materials used in structural beams in *Alternative 1* the algorithm took their lengths and multiplied them with the area of beam profile (or profiles) in that design option. This volume was then combined with LCA data of material used in that beam type. Finally, values for each type were summarised by beam types.

However, this method is subjective to a model preparation. It can be illustrated with the example of calculation of wall's environmental impact. In this case, wall's area was used for the volume calculation, with multiplication by material thicknesses in each option. That *area* in the concept model is calculated in one axis: wall's location line. Depending on its location in centre, exterior or interior face, one can get quite varying results. Nevertheless, it was accepted as satisfying the needs of an early LCA and conceptual model.

The process of comparison was further automated with direct export from Dynamo to Excel, and illustrated with dynamic charts to Power BI. Summary of results can be seen below. This allowed for relatively fast comprehension and comparison of performance for each alternative.

Energy Use

The analysis of expected energy use intensity was calculated with built-in *Autodesk Revit* and *Insight* tools. It was based on such information as: location, climate, occupancy of rooms and spaces, building volume and geometry, operational scheme, thermal properties of building elements, target temperatures, shading, mechanical systems.

In this case, it was necessary to apply significant modifications to the model, such as applying compound structures, which resulted in five different files. Values of thermal resistance for building elements were set accordingly to Polish building code, as required for 2021 standard for energy saving and thermal insulation[6]. The calculation was then set to analyse model with *Insight* software, and provided results for minimum, medium and maximum energy use, depending on the building operation.

Figure 9. Table of LCA results (in columns) for analysed material alternatives (in rows)

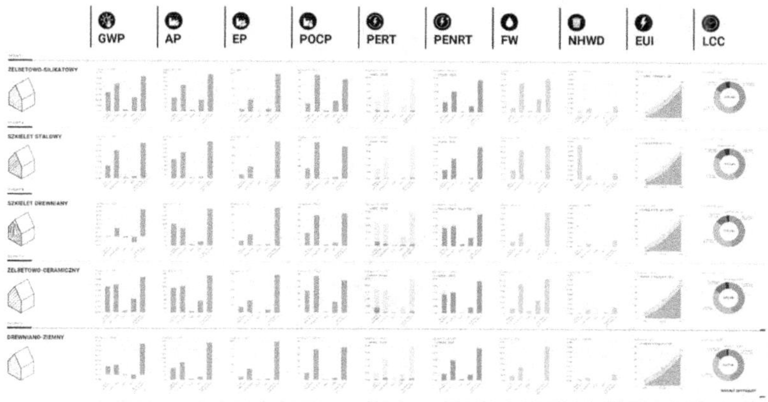

Cost Analysis

The cost analysis was performed on the basis of quantity take-offs from the BIM model, combined with information concerning prices in the building industry from SEKOCENBUD – local data company. Data from the schedules was connected with Dynamo to the Excel schedule, where all necessary prices were collected and totals for each design alternative calculated.

Contrary to the LCA analysis, LCC contains all phases of building's life cycle: construction, use stage (maintenance and use), and end of life. Using the expected life of implemented products, repairs and replacements were also included. The calculated building life cycle period was 30 years, with hypothetical demolition after that time.

Results And Discussion

The results were calculated for each of the five design alternatives in seventeen impact categories: twelve from EPD, one for EUI, and four for LCC. Each criteria was presented in separate chart, as in the Fig. In addition to comparative analysis, some categories were weighted as more important than the others; amongst them were carbon footprint, use of fresh water, disposed waste, energy use intensity and operational costs. Weighted assessment is presented in the Fig.34.

Figure 10a. Sample spider charts presenting calculated environmental impact for each design alternative. Global warming potential

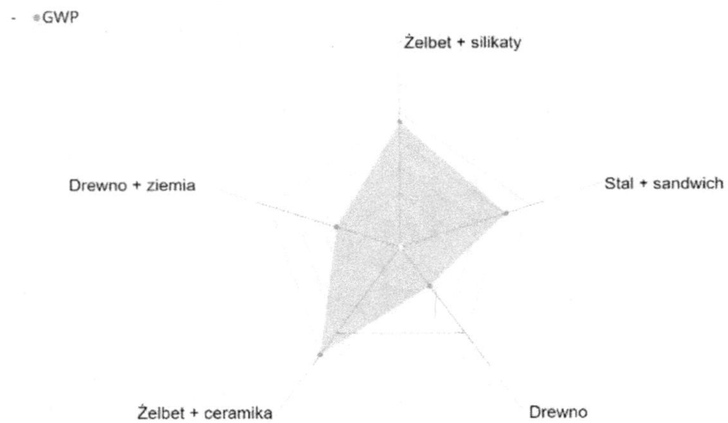

Figure 10b. Hazardous waste disposed

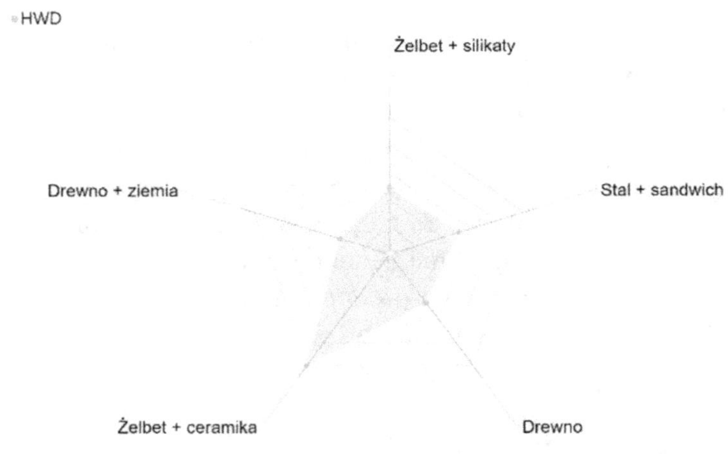

Within selected design options the most sustainable option was alternative 5 – Rammed earth with timber structure. It's impacts were significantly lower than other alternatives in the greatest number of categories. On the contrary, the alternative with the highest burden was alternative 4 – Masonry with concrete structure. It's global warming potential and water use was comparable with alternative 1, but here also energy use and the total cost of maintenance were the highest.

Figure 10c. Primary energy total - renewable

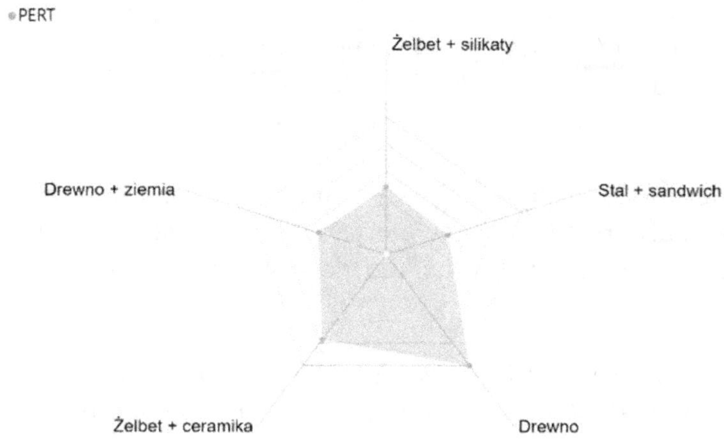

Figure 10d. Primary energy total - non-renewable

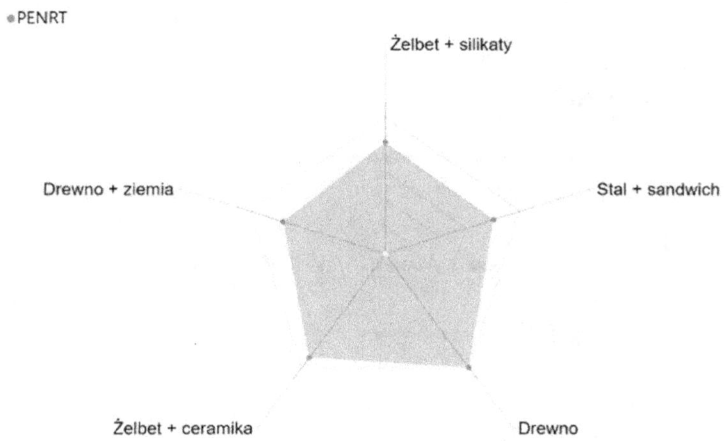

Moreover, rammed-earth alternative also appeared to be the most cost-effective in the long run. In the scope of 30 years both alternatives 1 and 4 with plaster finishes would require much more investment in preservation than rough rammed earth walls. Also steel sandwich panels in alternative 2 would need a sooner replacement.

To present what could have been achieved by selecting the most sustainable alternative the calculation of equivalent emissions to daily objects was included. Therefore, comparing the best and the worst alternative, one could save or prevent:

- emissions of GHG equivalent to the amount produced by 208 cars used for a year,
- enough energy to supply 131 homes for a year,
- volume of water that could fill 44 Olympic pools,
- disposal of waste equal to 73 dumpsters.

Presented study explored the way that architects could investigate more environmentally friendly design alternatives in a relatively fast assessment method. One observation was that the most time-consuming part of this task

Figure 11. Chart representation of decision matrix. The less the value, the less the environmental burden (blue and yellow), energy use (red) and cost (grey).

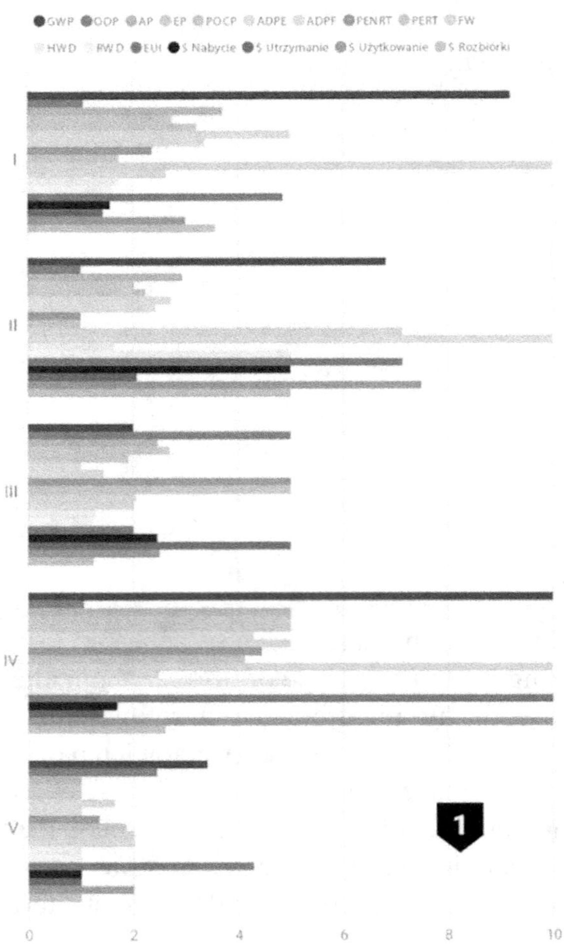

was data collection. There are many various ways to achieve optimisation. One of them could be an implementation of publicly available library of EPDs, linkable to a BIM model. For example, using computational tools such as Dynamo, EPD values from spreadsheet with catalogue of materials could be easily linked to compute the calculations, or update them when design elements change.

Advancement in this research would be further investigation of building's structure and form. Once set in this example, building's structural grid and size of elements was not optimised for lean construction, as it was not in the scope of study. In further works, this area of investigation could have a meaningful impact on the change of every design option performance. While

Figure 12. Illustration of potential savings, comparing the least and the most impactful design option.

iterating between subsequent options, the use of automated LCA study would be a great advantage.

CONCLUSION

BIM-based LCA examination proves that architect's choices have a tremendous impact on a building environmental footprint. Once the model is set, it can be investigated in many categories of impact, such as presented sustainability, life-cycle cost and energy use. The use of computational tools greatly enhances the opportunity to investigate plenty of design options with a minimum additional effort to the initial BIM model. This way, an architect can be focused on shaping the main vision of a project, and still have the chance to quickly verify the life cycle impact of proposed solutions.

Provided workflow can be used for initial stages of design, while assessing different design alternatives within the same geometry, for a conceptual model. It provides overall results of used building materials, showing the general idea of building's impact. This kind of analysis can be a useful guide to which material composition would be the best fit within given scope.

Ewelina Szczuka
Grupa 5 Architekci, Poland

***Ewelina Szczuka** is an architect, currently working at Grupa 5 Architekci office in Warsaw. She graduated from Warsaw University of Technology, where she received a master's degree in architecture. She gained experience in BIM methodology as well as the subject of a building's life cycle while working on the AEC Global Teamwork project, at PBL Lab, Stanford University. Earlier, she received architectural training at Strathclyde University in Glasgow and Delft University of Technology during student exchange programmes. She is interested in sustainable development, natural environment, and how contemporary techniques - data driven design, agile approach, automation - can help us advance our design.*

REFERENCES

Dz.U.2002.0690. (2002). *Rozporządzenie Ministra Infrastrktury Z Dnia 12 Kwietnia 2002 R. W Sprawie Warunków Technicznych, Jakim Powinny Odpowiadać Budynki I Ich Usytuowanie.*

European Commission. (2016). *Construction and Demolition Waste – Environment*. Retrieved from http://ec.europa.eu/environment/waste/construction_demolition.htm

IBU. (2018). *Published EPDs | Institut Bauen Und Umwelt e.V.* Retrieved from https://ibu-epd.com/en/published-epds/

ITB. (2018). *EPD | ITB*. Retrieved from http://www.itb.pl/epd.html

Polski Komitet Normalizacyjny. (2014). *EN 15804, Sustainability of Construction Works - Environmental Product Declarations.*

Polski Komitet Normalizacyjny. (2009). EN Zarządzanie Środowiskowe. Ocena Cyklu Życia. Zasady I Struktura. Warszawa: Polski Komitet Normalizacyjny.

ENDNOTES

1 European Commission, 'Construction and Demolition Waste - Environment', *Ec.europa.eu*, 2016 <http://ec.europa.eu/environment/waste/construction_demolition.htm> [accessed 13 June 2018].

2 Polski Komitet Normalizacyjny, *EN 15804, Sustainability of Construction Works - Environmental Product Declarations*, 2014.

3 Polski Komitet Normalizacyjny, *Zarządzanie Środowiskowe. Ocena Cyklu Życia. Zasady I Struktura* (Warszawa: Polski Komitet Normalizacyjny, 2009).

4 IBU, 'Published EPDs | Institut Bauen Und Umwelt e.V.', Www.ibu-Epd.com, 2018 <https://ibu-epd.com/en/published-epds/> [accessed 21 August 2018].

5 ITB, 'EPD | ITB', Www.itb.pl, 2018 <http://www.itb.pl/epd.html> [accessed 21 August 2018].

6 Dz.U.2002.0690, 'Rozporządzenie Ministra Infrastrktury Z Dnia 12 Kwietnia 2002 R. W Sprawie Warunków Technicznych, Jakim Powinny Odpowiadać Budynki I Ich Usytuowanie', 2002.

Case Study 2: MARS: 10-50-150. Circular Economy of the Martian Architecture – A 3D-Printed Ice Habitat

Architecture of a Martian habitat. *Construction of the first Martian home is just a matter of time. Identification of problems characteristic for the Martian conditions and long term manned missions is required for proper designing which has been based on a OLTARIS simulations. Existing resources as used as the basic building material, also repeatable modular solutions and recirculation of water and waste. Master degree thesis, promotors: Rynska Elzbieta, and Joanna Klimowicz, PhD. Arch. and presented at the Faculty of Architecture, Warsaw University of Technology.*

ABSTRACT

For Earth, the exploration of Mars constitutes a necessity. The Red Planet is currently the only place beyond Earth capable of sustaining a new location for human civilisation. The science and technology gained from a Martian program can spur development back on Earth and potentially even help with our climate-change challenge. Humanity's journey to Mars has already started and the construction of a first Martian home is just a matter of time. Based on the principles of the circular economy, a MARS: 10-50-150 project has been proposed, exploiting the in-situ available materials and cost reduction through the commercially available transport systems.

Keywords:Mars, Earth, architecture, astronomy, physics, circular economy, habitat, base, colony, ICE environments, space, isolation, comfort, 3D-printing, ice, Galactic cosmic rays, GCR, solar particle event, SPE, ionizing radiation, OLTARIS, analogue mission, exploration, colonization, Tikhonravov Crater, Arabia Terra, parametric architecture, Grasshopper, Starship, SpaceX, BFR, optimization, context, ISRU

Figure 1. MARS: 10-50-150: first steps on the Martian surface and the construction of the Phase 2 settlement
Source: (author's study)

MARS: 10-50-150, Introduction

The design of the architectonic objects in ICE environments (ICE stands for Isolation, Confinement, Extreme) poses a bigger challenge than in friendly, earthly conditions. Especially on Mars – a failure to accommodate the Martian contexts can lead to serious consequences and even cause harm to the inhabitants.

MARS: 10-50-150 is a three-phase project for the establishment of a Martian base and the realisation of a crewed Mars program. It has been developed as the main part of the master's thesis project at the Warsaw University of Technology in 2019. The project focuses on the in-situ resource utilization, reduction of costs through the implementation of commercially available hardware and transport solutions, and the provision of comfortable, architectonic solutions to the characteristic problems of the Martian environment. The habitat has been designed in accordance with the main assumptions of the circular economy.

Conditions and Localization

Mars is the fourth planet in the Solar System. Its volume equals to 15% of the Earth's and its mass equals only to 11%. Due to the low gravity (0.38g) and the absence of magnetic field Mars can't sustain dense atmosphere. The average temperature on the surface oscillates around -63°C and the pressure around 636Pa. The distance from Earth to Mars, at the closest point equals 3.0 light minutes, at the furthest 22.3 light minutes. A flight to Mars can last from three to nine months. A few billion years ago Mars was covered with liquid water oceans, lakes and rivers. Now almost all the water has evaporated into space, leaving the planet dry and dead. The Martian environment is not friendly for humans. Temperatures are extremely low, it is impossible to function outside without a pressurised spacesuit. Surface is constantly bombarded by cosmic and solar radiation and the regolith[1] contains dangerous amounts of toxic perchlorates.

However, in comparison to all the other celestial bodies, as far as we presently know, Mars is the only planet capable of sustaining human life in a long term. The solar power is strong enough to support plants, the sol (Martian day – 24h39min) is very close to the day on Earth. Frozen water is relatively abundant, many other resources can be found in the Martian regolith and atmosphere. Life on Mars could thrive in closed, pressurised habitats running on loop management in various areas. The technologies developed for the needs of a Martian habitat (and later a colony) can potentially be used back on Earth to reduce the severity of climate change.

Five potential base localizations have been examined based on the water access, radiation levels, local contexts, temperatures and insolation. Out of them the Tikhonravov Crater has been selected as the best possible candidate. The site is abundant in water (10% by mass in the top 100cm layer of regolith), the radiation is blocked by the thick atmosphere (150mSv/y at the absolute height of -2000m), daily temperatures oscillate from -40°C to 0°C, the insolation is very high (up to 590W/m^2) and the interplanetary transport is easy thanks to the low latitude (low minimal orbital inclination of the starting spaceships).

Figure 2. Water content in the regolith by mass (Gamma Spectrometer, Mars Odyssey, 2001) and the dose equivalent values on the surface (Mars Radiation Experiment, Mars Odyssey, 2001). Out of the five proposed localizations "D" has been chosen: Tikhonravov Crater, Arabia Terra, 15°N 35°E
Source: (author's study)

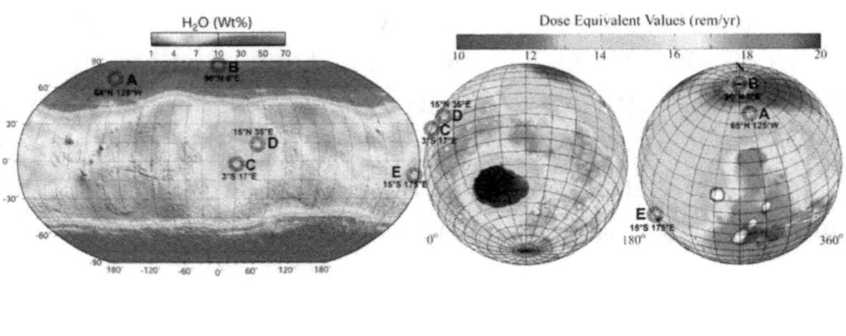

Construction Cycle, Phase 0, 06.2020-11.2026

Under the Martian conditions water ice is one of the best construction materials available. Clear ice is a relatively durable material (compressive strength of 10MPa, tensile strength of 1.2-2.5MPa, Petrovic, 2003), it transmits a lot of light (the extinction coefficient of the photo- synthetically active light equals 0.006cm^{-1}, Bolsenga 1978). Ice is a superior radiation shielding material that blocks all types of ionising radiation: galactic cosmic rays, solar particle events and radiation from the nuclear fuel used in the fission reactors. To assess the ice thickness needed to shield the inhabitants from various radiation sources a set of simulations has been conducted in the NASA On-Line Tool for the Assessment of Radiation in Space (OLTARIS) and a parametric algorithm has been developed, to determine the radiation levels inside of the modules. Ice structures can be easily manufactured using the 3D printing technology (Ice House, 2015). To protect the ice from sublimation the outer walls of the habitat are covered by a thin layer of glucose synthesized from the Martian atmosphere. The laminar flow air curtains are protecting the ice from melting on the inside.

Supplementary to the clear ice, two other materials can be 3D-printed on Mars. A 50:50 ice:regolith conglomerate has a higher compressive and tensile strength than clear Ice (Petrovic, 2003) and can thus be used for opaque walls and floor slabs. A material called duricrete is a Martian-regolith-based material developed by the Martin Marietta company in the 1980s. In its original form duricrete consisted just of the pre-wet regolith. After setting the material's strength equalled to 50% of the strength of concrete (Boyd, 1989). Duricrete can be strengthened by the addition of calcified gypsum cement (Kozicka,

Figure 3. The results of the conducted OLTARIS simulations. Ice thickness, light transmittance and galactic cosmic rays shielding (left, own study), ice thickness, light transmittance and solar particle event radiation (middle, personal study). Cold air curtains protecting the ice walls from melting (right, author's study)
Source: Left, own study. Middle, personal study. Right, author's study.

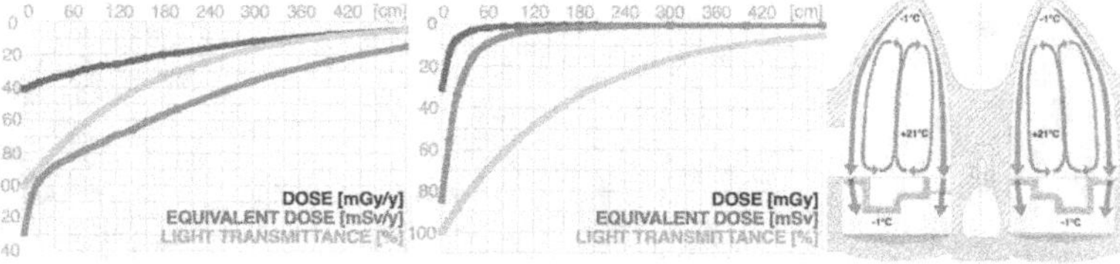

2008) and gravel, to bring its composition closer to the Earthly concrete. Both gypsum and gravel are abundant in the Martian regolith (Zubrin, 1997). The strengthened duricrete can be used for 3D-printing of partition walls, thin floor slabs, furniture and even simple tools and mechanisms. Duricrete doesn't melt in contact with the warm atmosphere inside of the habitat. For safety reasons the 3D-printing material must be treated with enzymes neutralizing the toxic perchlorates (2.5kg of enzymes per 1000kg of material, Davila, Willson, Coates, McKay, 2013).

Due to the pressure difference between the interior and the exterior of the base, the habitat's modules need to have a rounded geometry. A toroidal shape has been chosen, due to the highest possible construction span (compromised only by the pillar in the middle of the module). The parabolic sections of the arches are optimised for the 3D-printing process without the need for supports or formworks. The smooth, parabolic overhangs are self-supporting during the construction. Thanks to the toroidal sections of the modules, warm air can be trapped between the cold air-curtains preventing the walls from melting. The laminar flow curtains use slow and quiet air streams and thus they don't reduce the inhabitants' comfort.

The 3D-printing hardware is launched from Earth with a single SpaceX Starship-cargo rocket. A squad of small, autonomous robots runs the construction and conservation of the habitat. The robots are powered by the on-board radioisotope thermoelectric generators and the battery-stored energy from the habitat's solar panels, wind turbines and nuclear reactors. The whole ISRU system (In-situ resource utilization) needs 181-362kW of power. Each of the 20 3D-printing robots deposits 0.45-0.9m^3 of material per day. The whole construction of the 1st phase habitat takes 390-780 days It is fully automatic and doesn't need crewed supervision. After the habitat is ready the modules are tested under pressure to guarantee safety. When the tests are finalized, the 1st crew of 10 people is ready to fly to Mars.

Figure 4. Materials used in the construction of the habitat
Source: (author's study)

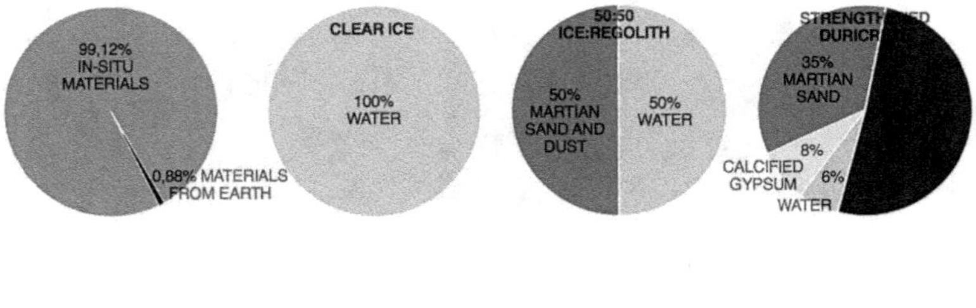

Figure 5. The proposed ISRU and 3D-printing mobile hardware. Small automatic excavators and 3D-printing robots. The robots can work horizontally or vertically thanks to the suction cups. Due to the low Martian pressure large, 82cm-wide cups are needed
Source: (author's study)

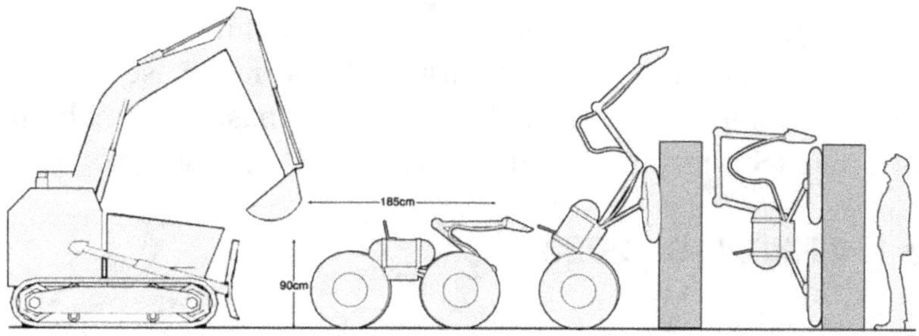

Figure 6a. Possible transit solutions (author's study)

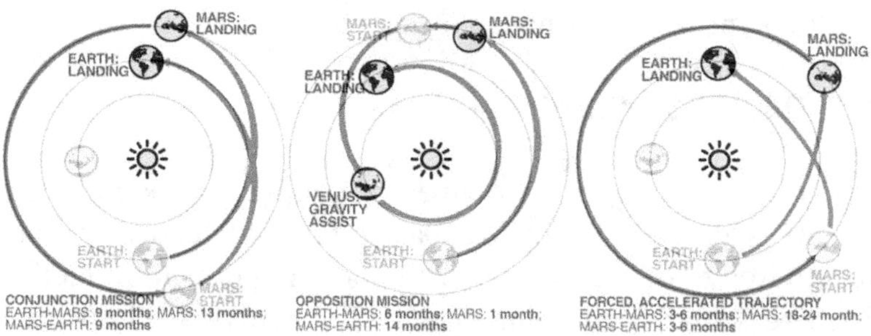

Figure 6b. commercially available transport systems
Source: (personal study based on the Starship rocket frame, SpaceX, 2017)

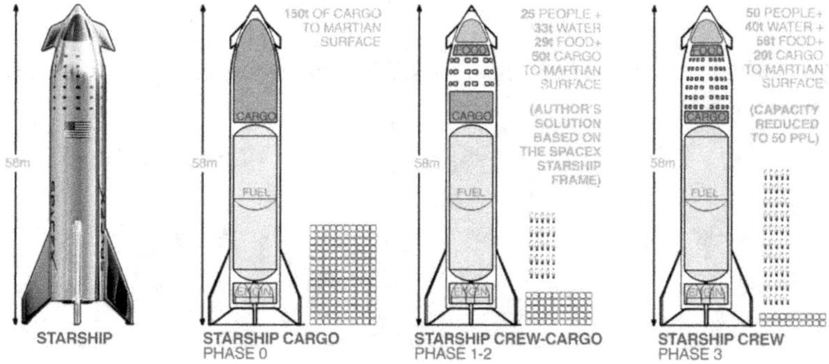

Phases

The project has been divided into three phases, each fulfilling different Mars exploration goals, and each designed for a different number of inhabitants:

Phase 0: Construction (Robotic surveys and 3D-printing)
Phase 1: Proving ground (Experimental and pioneering phase, 10 inhabitants)
Phase 2: Settlement (Scientific and exploration phase, 50 inhabitants
Phase 3: Town (Start of the Martian colony, 150+ inhabitants)

Phase 1, 11.2026-04.2033

The pioneering phase is the 1st step of humanity into the interplanetary exploration of Mars. During the three manned missions (10 crewmembers each) local surveys and research are conducted in the radius of 15-50km from the habitat. Martian crops reach their full capability, the construction of the phase 2 settlement and a permanent spaceport begins.

Phase 2, 04.2033-09.2039

While 10 is the most optimal number of people for long-duration space missions, 50 is enough to form a basic, family-friendly neighbourhood community. Growth of the Martian settlement requires introduction of new functions, like schools, entertainment modules, extended sports facilities and medical centres. Long-term science experiments are conducted, and long-range exploration starts (in the radius of 500km from the habitat). The light, 3D-printing equipment is supplemented by the heavy hardware. Local, valuable Martian resources are identified.

Figure 7. Diagrams of the MARS: 10-50-150 phases spanning from 2020 to 2043
Source: (author's study)

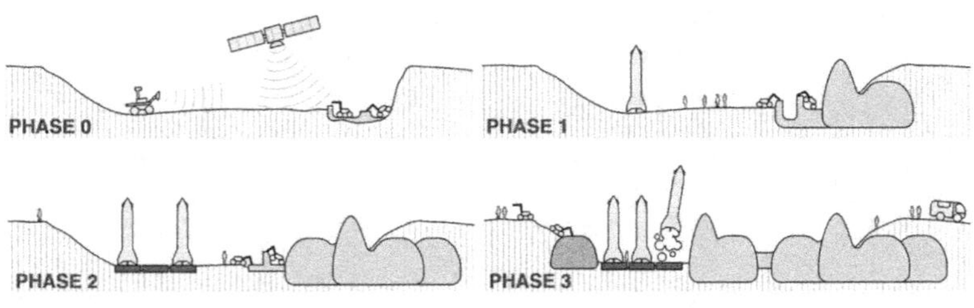

Figure 8. Section of the Phase 1 habitat
Source: (personal study)

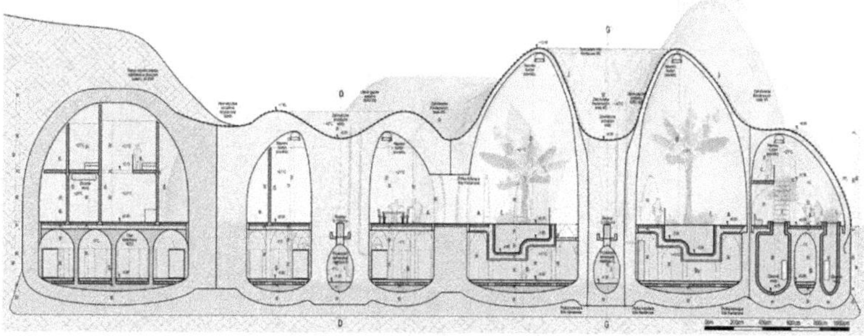

Phase 3, 09.2039-12.2043...

Permanent living in communities bigger than 50 people can lead to the feeling of alienation and can cause splits within the group. Separating the colony into smaller units of around 50-60 people tightens the personal bonds and helps the groups with forming their individual identities ("social groups" as defined by prof. J. M. Chmielewski, 2001). While spatially separated, the 50-people social units are connected by a fast-transit linear artery. During the 3rd phase the 1st Martian colony is officially established, and the first children are born on the Red Planet. Mars is explored in the global scale, the exploration of the Asteroid Belt starts, with Mars being its main hub. Due to the resource scarcity, the Martians develop the new fusion power solutions and CO_2 management techniques, that can be used back on Earth to limit the negative effects of the climate change.

Analog Base

In order to facilitate crew training on Earth an analogue base has been designed in the Bledowska Desert in Poland (50.3°N 19.5°E). Common training allows for the definition of the group's dynamics and helps in identification of the potential conflict situations. Both bases – on Mars and on Earth – are connected through a CCTV system allowing for a virtual presence of "Martians" on Earth and vice versa.

Figure 9. Phases 1-3 of the project, gradual growth from a 10-people habitat to a 150-people Martian town
Source: (author's study)

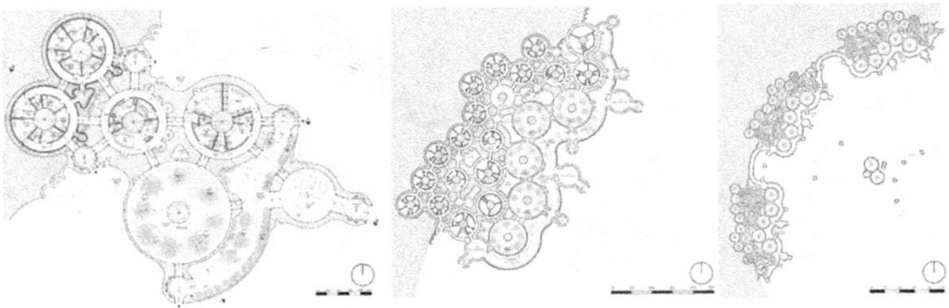

Figure 10. The MARS: 10-50-150 analogue base on Earth localized on the Bledowska Desert in Poland Source: (author's study)

Circular Aspects

The proposed MARS: 10-50-150 project has a potential of fulfilling some of the main goals required by the circular economy, as defined by the European Environment Agency. The circular aspects apply simultaneously to the Martian and earthly contexts, providing benefits for both planets and their inhabitants.

In the process of Mars exploration the interplanetary transport poses a major challenge and thus consumes most of the mission's budget. In-situ resource utilization and automatic, additive manufacturing of the habitat can minimize the required transport. In the construction Phase 0 only 70t of the ISRU and 3D-printing hardware must be transported from Earth to Mars. This equals to only 0.88% of the total mass of the used construction materials (7950t). 70t of cargo can be transported with a single flight of a SpaceX

Starship rocket (maximum capacity of 150t, SpaceX 2017). Reduction of the needed rocket launches results in a lower carbon footprint of the program. The small, 70t cargo doesn't depend on long and extensive industrial supply chains and thus doesn't put strain on the environment. The remaining 7880t of resources are available directly on the Martian surface in the form of regolith and atmosphere. In addition to the construction materials, the rocket fuel (methane and oxygen) for the transport system can additionally be easily synthesized on Mars. All that is needed is the atmospheric CO2 and water present just below the surface (up to 10% by mass).

The habitat systems exploit closed material and energetic loops. The ice construction's secondary function is water storage and buffer. The water circulates within the habitat in closed biological ecosystems supported by the supplementary environmental control and life support systems. Most of the food needed for the inhabitants is grown and bred locally. 22-25m2 of crop area per inhabitant provides up to 80% of total food supply. Waste is reduced through the extreme recycling practices. For example – food can be processed similarly to plastics allowing for the manufacturing of supply capsules and packages out of edible materials (Worf, 1964). These Earth-produced elements could then be used for preparation of the animal feed or fertilizers on Mars.

In the end of the life of the base no demolition is required. After the habitat is abandoned the ice structure will slowly evaporate over the course of few decades (slow sublimation of ice under Martian conditions). The water and Martian regolith will spontaneously return to the environment from which they were extracted. The duricrete elements will have the same mineralogical composition as the Martian rocks and will slowly erode in harmony with the surrounding landscape.

Figure 11. The resource cycles used in the construction and maintenance of the habitat
Source: (author's study)

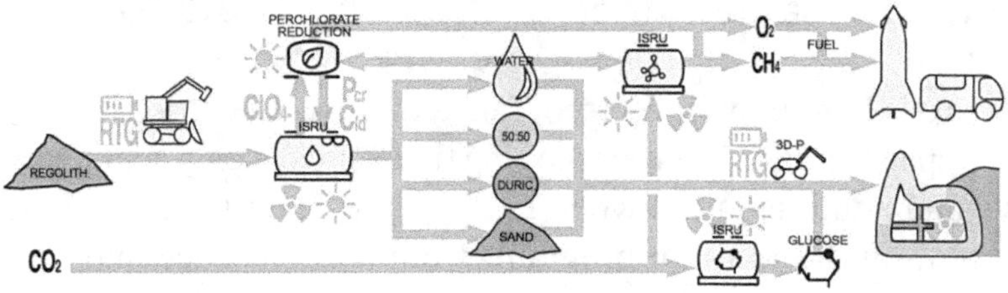

Costs and Energy Reduction in the Resource Retrieval Process

The resources needed for the construction are extracted and processed fully autonomously over long periods of time. The whole construction phase takes 390-780 days. Since no crewed supervision is needed, the construction pace can be slow resulting in a very high energy efficiency and low momentary power surges. All the materials are easily available and can be retrieved without severely impacting the local environment. Water ice is abundant in the top 100cm layer of loose regolith (Mars Odyssey, 2001), duricrete has the same mineralogical composition as the surrounding Martian rocks (Boyd, 1989). Glucose for the sublimation-protecting outer layer and useful polymer chemistry can be synthesized out of the atmospheric CO2 (NASA CO2 Conversion Challenge, 2019). During the maintenance the dirty glucose is stripped off the habitat's walls and replaced with a new layer. The old glucose can be used for the manufacturing of explosive materials (oxyliquits) essential for the geological exploration of Mars. The energetic system powering the ISRU processes is multimodal and maximizes the use of locally available sources. During a convenient weather solar panels are used. In case of a dust storm the loss of solar power is supplemented by the wind turbines (strong winds over 30m/s). Additionally, nuclear power is used in the form of small, 100kWe fission reactors and radioisotope thermoelectric generators. Nuclear fuel is the cheapest and most efficient power source that can be transported from Earth to Mars due to its high energy-density. In the short term the excess energy is saved in lithium-ion batteries and super-capacitors. In the long term the energy is stored chemically in the form of CH4 fuel and O2 oxidizer. Once the construction has started the potential losses in a case of failures and delays are minimal from the earthly point of view. All the possible material and energetic losses in the crew-less, construction phase don't affect the Earth and its economy.

Creation and Maintenance of the Local Work Place

Realisation of a Mars exploration program requires cooperation of many private and public space entities, each employing thousands of engineers, researchers, scientists and administration workers. It is important to remember, that the money spent on the Martian program is not sent to space and lost forever. Instead, it returns into the local economy where it's allowed to circulate.

The project's funds are channelled into the dynamic economic sectors that potentially can spur job creation, boost local economy and global growth, like: research and development, science and education (including STEM), innovative industries, space exploration, telecommunication, etc.

Creation of the Competition-Including Business Networks

The Space 4.0 movement, as defined by the ESA Technology Strategy (ESA, 2018), puts strong emphasis on the commercial aspect of space exploration. In the MARS: 10-50-150 project utilisation of commercially available transport systems and private hardware suppliers result in a huge cost reduction. Traditional, national space agencies have a very low decisive elasticity and tend to inflate the budgets of their projects. On the other hand, the private entities easily adapt to the changes and thanks to the commercial competition their solutions are cheaper even by few orders of magnitude.

Preservation and Stabilization of the Natural Environments

Finally, the establishment of a Martian colony can have profound effects on the climate change prevention back on Earth. The materials and power sources are scarce on Mars. The Martians will need efficient and innovative resource utilization and energetic systems. Fortunately – one of the most abundant Martian resources is carbon dioxide, which is the same compound that on Earth is responsible for most of the greenhouse effect. On Earth, there is no direct inducement for utilization of CO_2 as a resource, since there are easier,

Figure 12. Proposed interior design for the Starship Crew-Cargo
Source: (author's study based on the Starship rocket frame, SpaceX, 2017)

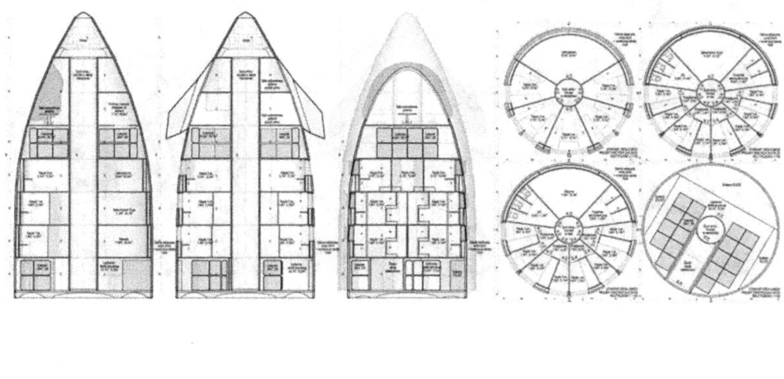

cheaper and historically conditioned options available. Thus, carbon dioxide is treated as a waste. On Mars however CO_2 can be used for the synthesis of methane fuel, sugars, polymers and other essential materials. Effective and cheap, Martian CO_2 processing techniques can be used back on Earth to reduce the carbon footprint of many of the industrial operations.

In return, the greenhouse effect processes accidentally unleashed on Earth and our new knowledge on the climate change can be used on Mars to kick-start the long process of terraforming. For example, huge carbon dioxide emissions caused by the heating of the southern polar cap in tandem with freon-emitting factories could rise the Martian atmospheric pressure up to 300hPa (Zubrin, 1997). This is enough for many types of plants and would be bearable for humans without the pressurised suits (only with scuba-diving-style oxygen masks).

Mars also provides a strong potential for the development of the new, clean energetic solutions. On Mars, deuterium (heavy isotope of hydrogen) is five times more abundant in water than on Earth (Zubrin, 1997). Deuterium is the main fuel for nuclear fusion – the same energy source that powers our Sun. The nuclear fusion consumes deuterium atoms and fuses them into helium producing huge amounts of energy in the process. Currently – the existing fusion reactors are very expensive and can't support fusion for longer periods of time. However, the Martian motivation comes into play again. With the very limited energy sources and high reserves of easily accessible fusion fuel Martians will be naturally inclined to develop new, clean, reliant and cheap fusion technologies. Once this is achieved the energetic crisis will end on Earth as well.

Figure 13. Mars 10-50-150 greenhouse module and the habitation module
Source: Left, personal study. Right, author's study.

CONCLUSION

The principles of the circular economy and circular architecture can be applied not only to the projects designed in the earthly contexts, but also for the Martian ones. Thanks to the use of local resources, closed material loops, commercial business networks it is possible to achieve a superior architectural quality of an ICE-environment habitat. What is more, the characteristic Martian contexts can potentially positively impact the environmental challenges back on Earth through the reduction of the carbon footprint of the industrial processes and the development of new, clean energy solutions.

Tomasz Dzieduszynski
Warsaw University of Technology, Poland

***Tomasz Dzieduszynski** is currently working on his doctoral thesis at the Warsaw University of Technology in the area of innovative, digital and multimedia tools for architecture. He holds an MSc degree in Architectural Engineering from the Warsaw University of Technology (Summa Cum Laude) and a BS degree in Architecture from the Silesian University of Technology. During his undergraduate studies, he spent a year as an Erasmus exchange student at the University of Coimbra, Portugal. In 2015, Tomasz was a finalist of the NASA 3D-printed habitat challenge. Tomasz has experience in interdisciplinary cooperation. He has participated in projects at the intersection of architecture, mechatronics, electronics, robotics, experimental literature, music and graphic design. His research interests include parametric architecture, sustainability and circular economy, BIM methodologies, space exploration and astronautics.*

REFERENCES

Bolsenga, S. J. (1979). Photosynthetically active radiation transmission through ice. NOAA Technical Memorandum,

Boyd, R., Thompson, P., & Clark, B. (1989). Duricrete and Composites Contruction on Mars, The Case for Mars III conference, San Diego,

Encyclopaedia Britannica. (1998). Regolith. In *The Encylopedia Britannica.*

Chmielewski, J. M. (2001). *Teoria Urbanistyki w projektowaniu i planowaniu miast*. Warsaw: Oficyna Wydawnicza Politechniki Warszawskiej.

Davila, A. F., Willson, J. D., Coates, J. D., & McKay, C. P. (2013). Coates, McKay, C. P.: Perchlorate on Mars: A chemical hazard and a resource for humans. *International Journal of Astrobiology*, *12*(4), 321–325. doi:10.1017/S1473550413000189

ESA. (2018). *ESA Technology Strategy Version 1.0*. ESA.

Kozicka, J. (2008). *Problemy architektoniczne bazy na Marsie jako habitatu w ekstremalnych warunkach*. Gdańsk University of Technology.

NASA. (2019). *Nasa CO_2 conversion challenge*. Retrieved from https://www.co2conversionchallenge.org/

Petrovic, J. J. (2003). Review Mechanical properties of ice and snow, *Journal of Materials Science*, (38), 1/

SEArch CloudsAO. (2019). Mars Ice House, 2015. Retrieved from http://www.marsicehouse.com/

Space X. (2017). Making life multiplanetary. In *68th International Astronautical Congress*, Adelaide, Australia

ENDNOTE

1 Regolith – loose unconsolidated rock and dust that sits atop a layer of bedrock (Encyclopaedia Britannica, 1998). Equivalent of earthly soil.

Related Readings

To continue IGI Global's long-standing tradition of advancing innovation through emerging research, please find below a compiled list of recommended IGI Global book chapters and journal articles in the areas of green cities, environmental management, and sustainable urban development. These related readings will provide additional information and guidance to further enrich your knowledge and assist you with your own research.

Ahuja, K., & Khosla, A. (2019). A Framework to Develop a Zero-Carbon Emission Sustainable Cognitive City. In K. Ahuja & A. Khosla (Eds.), *Driving the Development, Management, and Sustainability of Cognitive Cities* (pp. 1–26). Hershey, PA: IGI Global. doi:10.4018/978-1-5225-8085-0.ch001

Akkucuk, U. (2016). SCOR Model and the Green Supply Chain. In U. Akkucuk (Ed.), *Handbook of Research on Waste Management Techniques for Sustainability* (pp. 108–124). Hershey, PA: IGI Global. doi:10.4018/978-1-4666-9723-2.ch006

Albuquerque, C. M. (2017). Cities Really Smart and Inclusive: Possibilities and Limits for Social Inclusion and Participation. In L. Carvalho (Ed.), *Handbook of Research on Entrepreneurial Development and Innovation Within Smart Cities* (pp. 229–247). Hershey, PA: IGI Global. doi:10.4018/978-1-5225-1978-2.ch011

Angelucci, F., Di Girolamo, C., & Zazzero, E. (2018). New Designing Codes for Urban Infrastructures: A Hypothesis of a Transdisciplinary Approach. In G. Carlone, N. Martinelli, & F. Rotondo (Eds.), *Designing Grid Cities for Optimized Urban Development and Planning* (pp. 209–237). Hershey, PA: IGI Global. doi:10.4018/978-1-5225-3613-0.ch012

Avdimiotis, S., & Tilikidou, I. (2017). Smart Tourism Development: The Case of Halkidiki. In L. Carvalho (Ed.), *Handbook of Research on Entrepreneurial Development and Innovation Within Smart Cities* (pp. 491–513). Hershey, PA: IGI Global. doi:10.4018/978-1-5225-1978-2.ch021

Ayvaz, B., & Görener, A. (2016). Reverse Logistics in the Electronics Waste Industry. In U. Akkucuk (Ed.), *Handbook of Research on Waste Management Techniques for Sustainability* (pp. 155–171). Hershey, PA: IGI Global. doi:10.4018/978-1-4666-9723-2.ch008

Bagnato, V. P. (2018). Technology and Urban Structure: The Grid City Between Technological Innovation and New Public Space System. In G. Carlone, N. Martinelli, & F. Rotondo (Eds.), *Designing Grid Cities for Optimized Urban Development and Planning* (pp. 238–253). Hershey, PA: IGI Global. doi:10.4018/978-1-5225-3613-0.ch013

Bala, I., & Singh, G. (2019). Green Communication for Cognitive Cities. In K. Ahuja & A. Khosla (Eds.), *Driving the Development, Management, and Sustainability of Cognitive Cities* (pp. 87–110). Hershey, PA: IGI Global. doi:10.4018/978-1-5225-8085-0.ch004

Bernardino, S., & Santos, J. F. (2017). Building Smarter Cities through Social Entrepreneurship. In L. Carvalho (Ed.), *Handbook of Research on Entrepreneurial Development and Innovation Within Smart Cities* (pp. 327–362). Hershey, PA: IGI Global. doi:10.4018/978-1-5225-1978-2.ch015

Bernardo, M. D. (2017). Smart City Governance: From E-Government to Smart Governance. In L. Carvalho (Ed.), *Handbook of Research on Entrepreneurial Development and Innovation Within Smart Cities* (pp. 290–326). Hershey, PA: IGI Global. doi:10.4018/978-1-5225-1978-2.ch014

Bhatia, P., & Singh, P. (2019). Technological and Gamified Solutions for Pollution Control in Cognitive Cities. In K. Ahuja & A. Khosla (Eds.), *Driving the Development, Management, and Sustainability of Cognitive Cities* (pp. 234–249). Hershey, PA: IGI Global. doi:10.4018/978-1-5225-8085-0.ch010

Bhatt, R. (2017). Zero Tillage for Mitigating Global Warming Consequences and Improving Livelihoods in South Asia. In W. Ganpat & W. Isaac (Eds.), *Environmental Sustainability and Climate Change Adaptation Strategies* (pp. 126–161). Hershey, PA: IGI Global. doi:10.4018/978-1-5225-1607-1.ch005

Bolívar, M. P. (2018). Governance in Smart Cities: A Comparison of Practitioners' Perceptions and Prior Research. *International Journal of E-Planning Research*, *7*(2), 1–19. doi:10.4018/IJEPR.2018040101

Bostanci, S. H., & Albayrak, A. N. (2017). The Role of Eco-Municipalities in Climate Change for a Sustainable Future. In W. Ganpat & W. Isaac (Eds.), *Environmental Sustainability and Climate Change Adaptation Strategies* (pp. 213–231). Hershey, PA: IGI Global. doi:10.4018/978-1-5225-1607-1.ch008

Brkljačić, T., Majetić, F., & Tarabić, B. N. (2017). Smart Environment: Cyber Parks (Connecting Nature and Technology). In L. Carvalho (Ed.), *Handbook of Research on Entrepreneurial Development and Innovation Within Smart Cities* (pp. 150–172). Hershey, PA: IGI Global. doi:10.4018/978-1-5225-1978-2.ch008

Cai, Y. Z., Wu, F., Li, J., Wang, J., & Huang, M. (2019). Application of UAV Technology to Planning Study on Chinese Villages in Guanzhong. In J. Vargas-Hernández & J. Zdunek-Wielgołaska (Eds.), *Bioeconomical Solutions and Investments in Sustainable City Development* (pp. 180–195). Hershey, PA: IGI Global. doi:10.4018/978-1-5225-7958-8.ch008

Calisto, M. D., & Gonçalves, A. (2017). Smart Citizens, Wise Decisions: Sustainability-Driven Tourism Entrepreneurs. In L. Carvalho (Ed.), *Handbook of Research on Entrepreneurial Development and Innovation Within Smart Cities* (pp. 20–43). Hershey, PA: IGI Global. doi:10.4018/978-1-5225-1978-2.ch002

Capelo, C. (2017). Exploring the Dynamics of an Energy Service Venture. In L. Carvalho (Ed.), *Handbook of Research on Entrepreneurial Development and Innovation Within Smart Cities* (pp. 269–289). Hershey, PA: IGI Global. doi:10.4018/978-1-5225-1978-2.ch013

Caridi, G. (2018). Calabria 1783: The Orthogonal Grid as a Physical and Ideological Device of Reconstruction. In G. Carlone, N. Martinelli, & F. Rotondo (Eds.), *Designing Grid Cities for Optimized Urban Development and Planning* (pp. 176–187). Hershey, PA: IGI Global. doi:10.4018/978-1-5225-3613-0.ch010

Carlone, G. (2018). Cities and Extension Plans in the Kingdom of the Two Sicilies: Borgo Murattiano of Bari (1812-1859). In G. Carlone, N. Martinelli, & F. Rotondo (Eds.), *Designing Grid Cities for Optimized Urban Development and Planning* (pp. 1–18). Hershey, PA: IGI Global. doi:10.4018/978-1-5225-3613-0.ch001

Carvalho, L. C. (2017). Entrepreneurial Ecosystems: Lisbon as a Smart Start-Up City. In L. Carvalho (Ed.), *Handbook of Research on Entrepreneurial Development and Innovation Within Smart Cities* (pp. 1–19). Hershey, PA: IGI Global. doi:10.4018/978-1-5225-1978-2.ch001

Castagnolo, V. (2018). Analyzing, Classifying, Safeguarding: Drawing for the Borgo Murattiano Neighbourhood of Bari. In G. Carlone, N. Martinelli, & F. Rotondo (Eds.), *Designing Grid Cities for Optimized Urban Development and Planning* (pp. 93–108). Hershey, PA: IGI Global. doi:10.4018/978-1-5225-3613-0.ch006

Castilla-Polo, F., Gallardo-Vázquez, D., Sánchez-Hernández, M. I., & Ruiz-Rodríguez, M. D. (2017). Cooperatives as Responsible and Innovative Entrepreneurial Ecosystems in Smart Territories: The Olive Oil Industry in the South of Spain. In L. Carvalho (Ed.), *Handbook of Research on Entrepreneurial Development and Innovation Within Smart Cities* (pp. 459–490). Hershey, PA: IGI Global. doi:10.4018/978-1-5225-1978-2.ch020

Charalabidis, Y., & Theocharopoulou, C. (2019). A Participative Method for Prioritizing Smart City Interventions in Medium-Sized Municipalities. *International Journal of Public Administration in the Digital Age*, *6*(1), 41–63. doi:10.4018/IJPADA.2019010103

Chatsiwa, J., Mujere, N., & Maiyana, A. B. (2016). Municipal Solid Waste Management. In U. Akkucuk (Ed.), *Handbook of Research on Waste Management Techniques for Sustainability* (pp. 19–43). Hershey, PA: IGI Global. doi:10.4018/978-1-4666-9723-2.ch002

Chawla, R., Singhal, P., & Garg, A. K. (2019). Impact of Dust for Solar PV in Indian Scenario: Experimental Analysis. In K. Ahuja & A. Khosla (Eds.), *Driving the Development, Management, and Sustainability of Cognitive Cities* (pp. 111–138). Hershey, PA: IGI Global. doi:10.4018/978-1-5225-8085-0.ch005

Cılız, N., Yıldırım, H., & Temizel, Ş. (2016). Structure Development for Effective Medical Waste and Hazardous Waste Management System. In U. Akkucuk (Ed.), *Handbook of Research on Waste Management Techniques for Sustainability* (pp. 303–327). Hershey, PA: IGI Global. doi:10.4018/978-1-4666-9723-2.ch016

Collignon de Alba, C., Haberleithner, J., & López, M. M. (2017). Creative Industries in the Smart City: Overview of a Liability in Emerging Economies. In L. Carvalho (Ed.), *Handbook of Research on Entrepreneurial Development and Innovation Within Smart Cities* (pp. 107–126). Hershey, PA: IGI Global. doi:10.4018/978-1-5225-1978-2.ch006

Corum, A. (2016). Remanufacturing, an Added Value Product Recovery Strategy. In U. Akkucuk (Ed.), *Handbook of Research on Waste Management Techniques for Sustainability* (pp. 347–367). Hershey, PA: IGI Global. doi:10.4018/978-1-4666-9723-2.ch018

Crosas, C. (2018). Latin American Cities: Modern Grids From 1850s. In G. Carlone, N. Martinelli, & F. Rotondo (Eds.), *Designing Grid Cities for Optimized Urban Development and Planning* (pp. 39–51). Hershey, PA: IGI Global. doi:10.4018/978-1-5225-3613-0.ch003

D'Onofrio, S., Habenstein, A., & Portmann, E. (2019). Ontological Design for Cognitive Cities: The New Principle for Future Urban Management. In K. Ahuja & A. Khosla (Eds.), *Driving the Development, Management, and Sustainability of Cognitive Cities* (pp. 183–211). Hershey, PA: IGI Global. doi:10.4018/978-1-5225-8085-0.ch008

Dalzero, S. (2018). The Time of the Finished World Has Begun: A New Map of the World – National Borders Partially or Fully Fenced-Off. In G. Carlone, N. Martinelli, & F. Rotondo (Eds.), *Designing Grid Cities for Optimized Urban Development and Planning* (pp. 254–275). Hershey, PA: IGI Global. doi:10.4018/978-1-5225-3613-0.ch014

Damurski, L. (2016). Recent Progress in Online Communication Tools for Urban Planning: A Comparative Study of Polish and German Municipalities. *International Journal of E-Planning Research, 5*(1), 39–54. doi:10.4018/IJEPR.2016010103

Damurski, L. (2016). Smart City, Integrated Planning, and Multilevel Governance: A Conceptual Framework for e-Planning in Europe. *International Journal of E-Planning Research, 5*(4), 41–53. doi:10.4018/IJEPR.2016100103

Das, K. K., & Sharma, N. K. (2016). Post Disaster Housing Management for Sustainable Urban Development: A Review. *International Journal of Geotechnical Earthquake Engineering*, *7*(1), 1–18. doi:10.4018/IJGEE.2016010101

Das, S., & Nayyar, A. (2019). Innovative Ideas to Manage Urban Traffic Congestion in Cognitive Cities. In K. Ahuja & A. Khosla (Eds.), *Driving the Development, Management, and Sustainability of Cognitive Cities* (pp. 139–162). Hershey, PA: IGI Global. doi:10.4018/978-1-5225-8085-0.ch006

Deakin, M., & Reid, A. (2017). The Embedded Intelligence of Smart Cities: Urban Life, Citizenship, and Community. *International Journal of Public Administration in the Digital Age*, *4*(4), 62–74. doi:10.4018/IJPADA.2017100105

Djukic, A., Stupar, A., & Antonic, B. M. (2018). The Orthogonal Urban Matrix of the Towns in Vojvodina, Northern Serbia: Genesis and Transformation. In G. Carlone, N. Martinelli, & F. Rotondo (Eds.), *Designing Grid Cities for Optimized Urban Development and Planning* (pp. 128–156). Hershey, PA: IGI Global. doi:10.4018/978-1-5225-3613-0.ch008

Dolunay, O. (2016). A Paradigm Shift: Empowering Farmers to Eliminate the Waste in the Form of Fresh Water and Energy through the Implementation of 4R+T. In U. Akkucuk (Ed.), *Handbook of Research on Waste Management Techniques for Sustainability* (pp. 368–379). Hershey, PA: IGI Global. doi:10.4018/978-1-4666-9723-2.ch019

Dowling, C. M., Walsh, S. D., Purcell, S. M., Hynes, W. M., & Rhodes, M. L. (2017). Operationalising Sustainability within Smart Cities: Towards an Online Sustainability Indicator Tool. *International Journal of E-Planning Research*, *6*(4), 1–17. doi:10.4018/IJEPR.2017100101

Ekman, U. (2018). Smart City Planning: Complexity. *International Journal of E-Planning Research*, *7*(3), 1–21. doi:10.4018/IJEPR.2018070101

Erdogan, S. (2016). The Effect of Working Capital Management on Firm's Profitability: Evidence from Istanbul Stock Exchange. In U. Akkucuk (Ed.), *Handbook of Research on Waste Management Techniques for Sustainability* (pp. 244–261). Hershey, PA: IGI Global. doi:10.4018/978-1-4666-9723-2.ch013

Eudoxie, G., & Roopnarine, R. (2017). Climate Change Adaptation and Disaster Risk Management in the Caribbean. In W. Ganpat & W. Isaac (Eds.), *Environmental Sustainability and Climate Change Adaptation Strategies* (pp. 97–125). Hershey, PA: IGI Global. doi:10.4018/978-1-5225-1607-1.ch004

Fasolino, I. (2018). Rules for a New Town After a Disaster: The Gridded Schemes in the Plans. In G. Carlone, N. Martinelli, & F. Rotondo (Eds.), *Designing Grid Cities for Optimized Urban Development and Planning* (pp. 157–175). Hershey, PA: IGI Global. doi:10.4018/978-1-5225-3613-0.ch009

Fernandes, C. M., & Dias de Sousa, I. (2017). Digital Swarms: Social Interaction and Emergent Phenomena in Personal Communications Networks. In L. Carvalho (Ed.), *Handbook of Research on Entrepreneurial Development and Innovation Within Smart Cities* (pp. 44–59). Hershey, PA: IGI Global. doi:10.4018/978-1-5225-1978-2.ch003

Fernandes, V., Moreira, A., & Daniel, A. I. (2017). A Qualitative Analysis of Social Entrepreneurship Involving Social Innovation and Intervention. In L. Carvalho (Ed.), *Handbook of Research on Entrepreneurial Development and Innovation Within Smart Cities* (pp. 417–438). Hershey, PA: IGI Global. doi:10.4018/978-1-5225-1978-2.ch018

Gabriel, B. F., Valente, R. A., Dias-de-Oliveira, J., Neto, V. F., & Andrade-Campos, A. (2017). Methodologies for Engineering Learning and Teaching (MELT): An Overview of Engineering Education in Europe and a Novel Concept for Young Students. In L. Carvalho (Ed.), *Handbook of Research on Entrepreneurial Development and Innovation Within Smart Cities* (pp. 363–391). Hershey, PA: IGI Global. doi:10.4018/978-1-5225-1978-2.ch016

Garrick, T. A., & Liburd, O. E. (2017). Impact of Climate Change on a Key Agricultural Pest: Thrips. In W. Ganpat & W. Isaac (Eds.), *Environmental Sustainability and Climate Change Adaptation Strategies* (pp. 232–254). Hershey, PA: IGI Global. doi:10.4018/978-1-5225-1607-1.ch009

Gencer, Y. G. (2016). Mystery of Recycling: Glass and Aluminum Examples. In U. Akkucuk (Ed.), *Handbook of Research on Waste Management Techniques for Sustainability* (pp. 172–191). Hershey, PA: IGI Global. doi:10.4018/978-1-4666-9723-2.ch009

Gencer, Y. G., & Akkucuk, U. (2016). Reverse Logistics: Automobile Recalls and Other Conditions. In U. Akkucuk (Ed.), *Handbook of Research on Waste Management Techniques for Sustainability* (pp. 125–154). Hershey, PA: IGI Global. doi:10.4018/978-1-4666-9723-2.ch007

Gonçalves, J. M., Martins, T. G., & Vilhena da Cunha, I. B. (2017). Local Creative Ecosystems as a Strategy for the Development of Low-Density Urban Spaces. In L. Carvalho (Ed.), *Handbook of Research on Entrepreneurial Development and Innovation Within Smart Cities* (pp. 127–149). Hershey, PA: IGI Global. doi:10.4018/978-1-5225-1978-2.ch007

Goundar, S., & Appana, S. (2017). Mainstreaming Development Policies for Climate Change in Fiji: A Policy Gap Analysis and the Role of ICTs. In W. Ganpat & W. Isaac (Eds.), *Environmental Sustainability and Climate Change Adaptation Strategies* (pp. 1–31). Hershey, PA: IGI Global. doi:10.4018/978-1-5225-1607-1.ch001

Grochulska-Salak, M. (2019). Urban Farming in Sustainable City Development. In J. Vargas-Hernández & J. Zdunek-Wielgołaska (Eds.), *Bioeconomical Solutions and Investments in Sustainable City Development* (pp. 43–64). Hershey, PA: IGI Global. doi:10.4018/978-1-5225-7958-8.ch003

Gruszecka, K. (2019). Ecological Centre of Warsaw as a Development Path. In J. Vargas-Hernández & J. Zdunek-Wielgołaska (Eds.), *Bioeconomical Solutions and Investments in Sustainable City Development* (pp. 117–150). Hershey, PA: IGI Global. doi:10.4018/978-1-5225-7958-8.ch006

Hesapci-Sanaktekin, O., & Aslanbay, Y. (2016). The Networked Self: Collectivism Redefined in Civic Engagements through Social Media Causes. In U. Akkucuk (Ed.), *Handbook of Research on Waste Management Techniques for Sustainability* (pp. 262–276). Hershey, PA: IGI Global. doi:10.4018/978-1-4666-9723-2.ch014

Ikeda, M. (2019). Developing a Sustainable Eco-City in Pre-Olympic Tokyo: Potential of New Methods and Their Limits in an Urban Era. In J. Vargas-Hernández & J. Zdunek-Wielgołaska (Eds.), *Bioeconomical Solutions and Investments in Sustainable City Development* (pp. 196–223). Hershey, PA: IGI Global. doi:10.4018/978-1-5225-7958-8.ch009

İşcan, E. (2019). Strategies of Sustainable Bioeconomy in the Industry 4.0 Framework for Inclusive and Social Prosperity. In J. Vargas-Hernández & J. Zdunek-Wielgołaska (Eds.), *Bioeconomical Solutions and Investments in Sustainable City Development* (pp. 21–42). Hershey, PA: IGI Global. doi:10.4018/978-1-5225-7958-8.ch002

J., J., Samui, P., & Dixon, B. (2016). Determination of Rate of Medical Waste Generation Using RVM, MARS and MPMR. In U. Akkucuk (Ed.), *Handbook of Research on Waste Management Techniques for Sustainability* (pp. 1-18). Hershey, PA: IGI Global. doi:10.4018/978-1-4666-9723-2.ch001

Jiménez, R., Rodríguez, P. L., & Fernández, R. R. (2019). Green Spaces of the Metropolitan Area of Guadalajara. In J. Vargas-Hernández & J. Zdunek-Wielgołaska (Eds.), *Bioeconomical Solutions and Investments in Sustainable City Development* (pp. 151–179). Hershey, PA: IGI Global. doi:10.4018/978-1-5225-7958-8.ch007

Kais, S. M. (2017). Climate Change: Vulnerability and Resilience in Commercial Shrimp Aquaculture in Bangladesh. In W. Ganpat & W. Isaac (Eds.), *Environmental Sustainability and Climate Change Adaptation Strategies* (pp. 162–187). Hershey, PA: IGI Global. doi:10.4018/978-1-5225-1607-1.ch006

Khan, B. (2019). Bio-Economy: Visions, Strategies, and Policies. In J. Vargas-Hernández & J. Zdunek-Wielgołaska (Eds.), *Bioeconomical Solutions and Investments in Sustainable City Development* (pp. 1–20). Hershey, PA: IGI Global. doi:10.4018/978-1-5225-7958-8.ch001

Khanna, B. K. (2017). Indian National Strategy for Climate Change Adaptation and Mitigation. In W. Ganpat & W. Isaac (Eds.), *Environmental Sustainability and Climate Change Adaptation Strategies* (pp. 32–63). Hershey, PA: IGI Global. doi:10.4018/978-1-5225-1607-1.ch002

Khanna, B. K. (2017). Vulnerability of the Lakshadweep Coral Islands in India and Strategies for Mitigating Climate Change Impacts. In W. Ganpat & W. Isaac (Eds.), *Environmental Sustainability and Climate Change Adaptation Strategies* (pp. 64–96). Hershey, PA: IGI Global. doi:10.4018/978-1-5225-1607-1.ch003

Kocasoy, G. (2016). Economic Instruments for Sustainable Environmental Management. In U. Akkucuk (Ed.), *Handbook of Research on Waste Management Techniques for Sustainability* (pp. 192–211). Hershey, PA: IGI Global. doi:10.4018/978-1-4666-9723-2.ch010

Lallo, C. H., Smalling, S., Facey, A., & Hughes, M. (2017). The Impact of Climate Change on Small Ruminant Performance in Caribbean Communities. In W. Ganpat & W. Isaac (Eds.), *Environmental Sustainability and Climate Change Adaptation Strategies* (pp. 296–321). Hershey, PA: IGI Global. doi:10.4018/978-1-5225-1607-1.ch011

Li, Z., Wang, Y., & Chen, Q. (2017). Real-Time Monitoring of Intercity Passenger Flows Based on Big Data: A Decision Support Tool for Urban Sustainability. *International Journal of Strategic Decision Sciences*, *8*(4), 120–128. doi:10.4018/IJSDS.2017100106

Loi, N. K., Huyen, N. T., Tu, L. H., Tram, V. N., Liem, N. D., Dat, N. L., ... Minh, D. N. (2017). Sustainable Land Use and Watershed Management in Response to Climate Change Impacts: Case Study in Srepok Watershed, Central Highland of Vietnam. In W. Ganpat & W. Isaac (Eds.), *Environmental Sustainability and Climate Change Adaptation Strategies* (pp. 255–295). Hershey, PA: IGI Global. doi:10.4018/978-1-5225-1607-1.ch010

López-Arranz, M. A. (2017). The Role Corporate Social Responsibility Has in the Smart City Project in Spain. In L. Carvalho (Ed.), *Handbook of Research on Entrepreneurial Development and Innovation Within Smart Cities* (pp. 439–458). Hershey, PA: IGI Global. doi:10.4018/978-1-5225-1978-2.ch019

Lucas, M. R., Rego, C., Vieira, C., & Vieira, I. (2017). Proximity and Cooperation for Innovative Regional Development: The Case of the Science and Technology Park of Alentejo. In L. Carvalho (Ed.), *Handbook of Research on Entrepreneurial Development and Innovation Within Smart Cities* (pp. 199–228). Hershey, PA: IGI Global. doi:10.4018/978-1-5225-1978-2.ch010

Magrinho, A., Neves, J., & Silva, J. R. (2017). The Triple Helix Model: Evidence in the Internationalization of the Health Industry. In L. Carvalho (Ed.), *Handbook of Research on Entrepreneurial Development and Innovation Within Smart Cities* (pp. 60–79). Hershey, PA: IGI Global. doi:10.4018/978-1-5225-1978-2.ch004

Maiorano, A. C. (2018). Urban Fronts in Murattiano Neighbourhood of Bari: A Selective Survey of the Built Environment. In G. Carlone, N. Martinelli, & F. Rotondo (Eds.), *Designing Grid Cities for Optimized Urban Development and Planning* (pp. 78–92). Hershey, PA: IGI Global. doi:10.4018/978-1-5225-3613-0.ch005

Martinelli, N., & Mangialardi, G. (2018). Cities With Grid Layout: Ubiquitousness and Flexibility of an Urban Model. In G. Carlone, N. Martinelli, & F. Rotondo (Eds.), *Designing Grid Cities for Optimized Urban Development and Planning* (pp. 188–208). Hershey, PA: IGI Global. doi:10.4018/978-1-5225-3613-0.ch011

Mizutani, S., Liao, K., & Sasaki, T. G. (2019). Forest-River-Ocean Nexus-Based Education for Community Development: Aiming at Resilient Sustainable Society. In J. Vargas-Hernández & J. Zdunek-Wielgołaska (Eds.), *Bioeconomical Solutions and Investments in Sustainable City Development* (pp. 224–248). Hershey, PA: IGI Global. doi:10.4018/978-1-5225-7958-8.ch010

Moreira, A., & Ferreira, M. A. (2017). Strategic Challenges of the Portuguese Molds Industry: A Sectoral Innovation Perspective. In L. Carvalho (Ed.), *Handbook of Research on Entrepreneurial Development and Innovation Within Smart Cities* (pp. 534–560). Hershey, PA: IGI Global. doi:10.4018/978-1-5225-1978-2.ch023

Mujere, N., & Moyce, W. (2017). Climate Change Impacts on Surface Water Quality. In W. Ganpat & W. Isaac (Eds.), *Environmental Sustainability and Climate Change Adaptation Strategies* (pp. 322–340). Hershey, PA: IGI Global. doi:10.4018/978-1-5225-1607-1.ch012

Mundula, L., & Auci, S. (2017). Smartness, City Efficiency, and Entrepreneurship Milieu. In L. Carvalho (Ed.), *Handbook of Research on Entrepreneurial Development and Innovation Within Smart Cities* (pp. 173–198). Hershey, PA: IGI Global. doi:10.4018/978-1-5225-1978-2.ch009

Nath, R. (2019). Parametric Evaluation of Beam Deflection on Piezoelectric Material Using Implicit and Explicit Method Simulations: A Study in Energy Engineering. In J. Vargas-Hernández & J. Zdunek-Wielgołaska (Eds.), *Bioeconomical Solutions and Investments in Sustainable City Development* (pp. 65–87). Hershey, PA: IGI Global. doi:10.4018/978-1-5225-7958-8.ch004

Nayyar, A., Jain, R., Mahapatra, B., & Singh, A. (2019). Cyber Security Challenges for Smart Cities. In K. Ahuja & A. Khosla (Eds.), *Driving the Development, Management, and Sustainability of Cognitive Cities* (pp. 27–54). Hershey, PA: IGI Global. doi:10.4018/978-1-5225-8085-0.ch002

Odabasi, A., & Tiryaki, C. S. (2016). An Empirical Review of Long Term Electricity Demand Forecasts for Turkey. In U. Akkucuk (Ed.), *Handbook of Research on Waste Management Techniques for Sustainability* (pp. 227–243). Hershey, PA: IGI Global. doi:10.4018/978-1-4666-9723-2.ch012

Okay, E. (2016). Towards Smart Cities in Turkey?: Transitioning from Waste to Creative, Clean and Cheap Eco-Energy. In U. Akkucuk (Ed.), *Handbook of Research on Waste Management Techniques for Sustainability* (pp. 277–302). Hershey, PA: IGI Global. doi:10.4018/978-1-4666-9723-2.ch015

Pawlikowska-Piechotka, A., Łukasik, N., Ostrowska-Tryzno, A., & Sawicka, K. (2017). A Smart City Initiative: Urban Greens and Evaluation Method of the Sport and Recreation Potentials (SEM). In L. Carvalho (Ed.), *Handbook of Research on Entrepreneurial Development and Innovation Within Smart Cities* (pp. 561–583). Hershey, PA: IGI Global. doi:10.4018/978-1-5225-1978-2.ch024

Raisinghani, M. S., & Idemudia, E. C. (2016). Green Information Systems for Sustainability. In U. Akkucuk (Ed.), *Handbook of Research on Waste Management Techniques for Sustainability* (pp. 212–226). Hershey, PA: IGI Global. doi:10.4018/978-1-4666-9723-2.ch011

Rathee, D. S., Ahuja, K., & Hailu, T. (2019). Role of Electronics Devices for E-Health in Smart Cities. In K. Ahuja & A. Khosla (Eds.), *Driving the Development, Management, and Sustainability of Cognitive Cities* (pp. 212–233). Hershey, PA: IGI Global. doi:10.4018/978-1-5225-8085-0.ch009

Rotondo, F. (2018). The Grid Cities: Between Tradition and Innovation. In G. Carlone, N. Martinelli, & F. Rotondo (Eds.), *Designing Grid Cities for Optimized Urban Development and Planning* (pp. 109–127). Hershey, PA: IGI Global. doi:10.4018/978-1-5225-3613-0.ch007

Sánchez-Fernández, M. D., & Cardona, J. R. (2017). The Perception of the Effect of Tourism on the Local Community before the Ibiza Smart Island Project. In L. Carvalho (Ed.), *Handbook of Research on Entrepreneurial Development and Innovation Within Smart Cities* (pp. 392–416). Hershey, PA: IGI Global. doi:10.4018/978-1-5225-1978-2.ch017

Santos, B. (2017). Improving Urban Planning Information, Transparency and Participation in Public Administrations. *International Journal of E-Planning Research, 6*(4), 58–75. doi:10.4018/IJEPR.2017100104

Sardà, J. F. (2018). Cerdà/Barcelona/Eixample: 1855-2017 … A Work in Progress. In G. Carlone, N. Martinelli, & F. Rotondo (Eds.), *Designing Grid Cities for Optimized Urban Development and Planning* (pp. 19–38). Hershey, PA: IGI Global. doi:10.4018/978-1-5225-3613-0.ch002

Selvi, M. S. (2016). Physical Distribution Problems of Textile Companies in Turkey. In U. Akkucuk (Ed.), *Handbook of Research on Waste Management Techniques for Sustainability* (pp. 328–346). Hershey, PA: IGI Global. doi:10.4018/978-1-4666-9723-2.ch017

Signorile, N. (2018). Pride and Prejudice: The Murattiano-Modernism. In G. Carlone, N. Martinelli, & F. Rotondo (Eds.), *Designing Grid Cities for Optimized Urban Development and Planning* (pp. 52–77). Hershey, PA: IGI Global. doi:10.4018/978-1-5225-3613-0.ch004

Singh, G., Kapoor, R., & Khosla, A. K. (2019). Intelligent Anomaly Detection Video Surveillance Systems for Smart Cities. In K. Ahuja & A. Khosla (Eds.), *Driving the Development, Management, and Sustainability of Cognitive Cities* (pp. 163–182). Hershey, PA: IGI Global. doi:10.4018/978-1-5225-8085-0.ch007

Singh-Ackbarali, D., & Maharaj, R. (2017). Mini Livestock Ranching: Solution to Reducing the Carbon Footprint and Negative Environmental Impacts of Agriculture. In W. Ganpat & W. Isaac (Eds.), *Environmental Sustainability and Climate Change Adaptation Strategies* (pp. 188–212). Hershey, PA: IGI Global. doi:10.4018/978-1-5225-1607-1.ch007

Sood, N., Saini, I., Awasthi, T., Saini, M. K., Bhoriwal, P., & Kaur, T. (2019). Fog Removal Algorithms for Real-Time Video Footage in Smart Cities for Safe Driving. In K. Ahuja & A. Khosla (Eds.), *Driving the Development, Management, and Sustainability of Cognitive Cities* (pp. 55–86). Hershey, PA: IGI Global. doi:10.4018/978-1-5225-8085-0.ch003

Stone, R. J. (2017). Modelling the Frequency of Tropical Cyclones in the Lower Caribbean Region. In W. Ganpat & W. Isaac (Eds.), *Environmental Sustainability and Climate Change Adaptation Strategies* (pp. 341–349). Hershey, PA: IGI Global. doi:10.4018/978-1-5225-1607-1.ch013

Stratigea, A., Leka, A., & Panagiotopoulou, M. (2017). In Search of Indicators for Assessing Smart and Sustainable Cities and Communities' Performance. *International Journal of E-Planning Research, 6*(1), 43–73. doi:10.4018/IJEPR.2017010103

Touq, A. B., & Ijeh, A. (2018). Information Security and Ecosystems in Smart Cities: The Case of Dubai. *International Journal of Information Systems and Social Change, 9*(2), 28–43. doi:10.4018/IJISSC.2018040103

Üç, M., & Elitaş, C. (2016). Life Cycle Costing for Sustainability. In U. Akkucuk (Ed.), *Handbook of Research on Waste Management Techniques for Sustainability* (pp. 96–107). Hershey, PA: IGI Global. doi:10.4018/978-1-4666-9723-2.ch005

Ulker-Demirel, E., & Demirel, E. (2016). Green Marketing and Stakeholder Perceptions. In U. Akkucuk (Ed.), *Handbook of Research on Waste Management Techniques for Sustainability* (pp. 75–95). Hershey, PA: IGI Global. doi:10.4018/978-1-4666-9723-2.ch004

Vargas-Hernández, J. G., López, J. J., & Zdunek-Wielgołaska, J. A. (2019). Entrepreneurial and Institutional Analysis of Biodiesel Companies in Mexico. In J. Vargas-Hernández & J. Zdunek-Wielgołaska (Eds.), *Bioeconomical Solutions and Investments in Sustainable City Development* (pp. 89–115). Hershey, PA: IGI Global. doi:10.4018/978-1-5225-7958-8.ch005

Vázquez, D. G., & Gil, M. T. (2017). Sustainability in Smart Cities: The Case of Vitoria-Gasteiz (Spain) – A Commitment to a New Urban Paradigm. In L. Carvalho (Ed.), *Handbook of Research on Entrepreneurial Development and Innovation Within Smart Cities* (pp. 248–268). Hershey, PA: IGI Global. doi:10.4018/978-1-5225-1978-2.ch012

Velozo, R. A., & Montanha, G. K. (2017). Evaluation of a Mobile Software Development Company. In L. Carvalho (Ed.), *Handbook of Research on Entrepreneurial Development and Innovation Within Smart Cities* (pp. 514–533). Hershey, PA: IGI Global. doi:10.4018/978-1-5225-1978-2.ch022

About the Author

Elzbieta Rynska works at Faculty of Architecture WUT since 1987. A title of a full professor since 2015. One of the initiators and coordinator of a 2 semester interdisciplinary course Architecture and Urban Planning in Sustainable Development, Master Level studies, at Faculty of Architecture Warsaw University of Technology. Author of various scientific books in Polish and English. Promoter of circa 40 Master and Bachelor Degree Diplomas and 6 PhD dissemination. Main interests: sustainable architecture and urban planning, interdisciplinary design processes, resilient and circular issues in city development, user health and well-being, energy efficient design and alternative energy technologies.

Index

Purchase Print, E-Book, or Print + E-Book

IGI Global's reference books can now be purchased from three unique pricing formats: Print Only, E-Book Only, or Print + E-Book.
Shipping fees may apply.

www.igi-global.com

Recommended Reference Books

ISBN: 978-1-5225-7727-0
© 2019; 335 pp.
List Price: $195

ISBN: 978-1-5225-5577-3
© 2019; 291 pp.
List Price: $195

ISBN: 978-1-5225-5291-8
© 2019; 252 pp.
List Price: $175

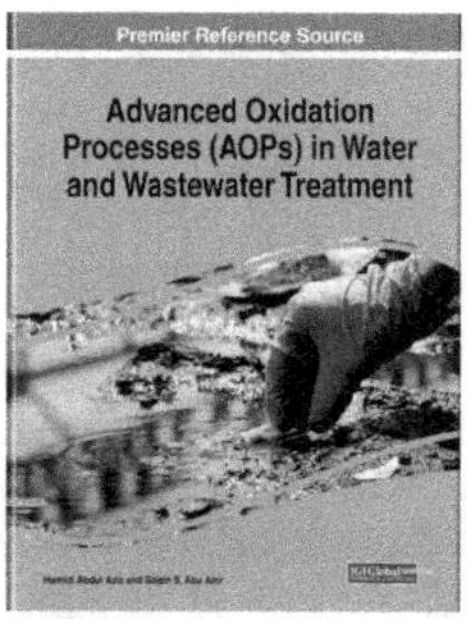

ISBN: 978-1-5225-5766-1
© 2019; 501 pp.
List Price: $195

ISBN: 978-1-5225-5909-2
© 2019; 308 pp.
List Price: $195

ISBN: 978-1-5225-8027-0
© 2019; 337 pp.
List Price: $205

Looking for free content, product updates, news, and special offers?
Join IGI Global's mailing list today and start enjoying exclusive perks sent only to IGI Global members.
Add your name to the list at **www.igi-global.com/newsletters.**

Publisher of Peer-Reviewed, Timely, and Innovative Academic Research

IGI Global
DISSEMINATOR OF KNOWLEDGE

www.igi-global.com Sign up at www.igi-global.com/newsletters facebook.com/igiglobal twitter.com/igiglobal

www.ingramcontent.com/pod-product-compliance
Lightning Source LLC
LaVergne TN
LVHW080048210926
848481LV00011B/3303
* 9 7 8 1 7 9 9 8 1 8 8 7 8 *